Student's Solution Manual:

Introduction to Basic Concepts in Engineering

for adept high school students

To accompany the 1st Edition text

© 2016 Dr. Andrew S. Heintz

TABLE OF CONTENTS

SOLUTIONS TO PRACTICE PROBLEMS

Chapter 4 – Units and Conversions
Chapter 5 – Electrical Circuits
Chapter 6 – Thermodynamics
Chapter 7 – Fluid Statics and Fluid Dynamics
Chapter 8 – Material and Energy Balances
Chapter 9 – Engineering Statistics
Chapter 10 – Computer Engineering
Chapter 11 – Reliability Engineering
Chapter 12 – Materials Science and Engineering
Chapter 13 – Industrial Manufacturing and Operations

SOLUTIONS TO COMPREHENSIVE LEARNING PROBLEMS

Chapter 5 – Resistance of a Bolt
Chapter 6 – Pot on a Stove
Chapter 7 – Water Tower Pump
Chapter 8 – Heat Exchanger
Chapter 12 – Bolt Strength
Chapter 13 – Internet Service Provider

Chapter 4 Solutions
UNITS AND CONVERSIONS

Problem 4.1
Convert 1500 grams into pounds.

$$\frac{1500\,g}{} \left|\frac{kg}{1000\,g}\right| \frac{2.20\,lb}{kg}$$

= 3.3 lb

Problem 4.2
A beaker contains 200mL. What is its volume in gallons?

$$\frac{200\,mL}{} \left|\frac{L}{1000\,mL}\right| \frac{1\,gal}{3.785\,L}$$

= 0.757 gal

Problem 4.3
Your car has a tank that fits 16 gallons of gas. How many barrels (bbls) of gasoline is this?

$$\frac{16\,gal}{} \left|\frac{bbl}{55\,gal}\right.$$

= 0.29 bbl

Problem 4.4
If a car is moving 55 mph, what is its speed in nm per second?

$$\frac{55\,mi}{hr} \left|\frac{hr}{3600\,s}\right| \frac{1609.34\,m}{mi} \left|\frac{10^9\,nm}{m}\right.$$

= 2.45 × 10^{10} nm/s

Problem 4.5
Determine the conversion factor between miles/hour and km/min.

~~mi~~	~~hr~~	1.609 km
~~hr~~	60 min	~~mi~~

= 0.0268

~~km~~	60 ~~min~~	mi
~~min~~	hr	1.609 ~~km~~

= 37.3

- mi/hr → km/min = approx. 0.0268
- km/min → mi/hr = approx. 37.3

Problem 4.6
Determine the conversion factor between g/mL and lb/in³.

g	~~mL~~	2.54 ~~cm~~	2.54 ~~cm~~	2.54 ~~cm~~	kg	2.20 lb
~~mL~~	~~cm³~~	in	in	in	1000 ~~g~~	~~kg~~

= 0.0268

~~lb~~	in	in	in	~~cm³~~	kg	1000 g
in³	2.54 ~~cm~~	2.54 ~~cm~~	2.54 ~~cm~~	~~mL~~	2.20 lb	~~kg~~

= 0.036

g/mL → lb/in³ = approx. 0.036
km/min → mi/hr = approx. 27.7

Problem 4.7
A box measures 3.12 ft in length, 0.0455 yd in width, and 7.87 inches in height. What is its volume in cubic centimeters?

Start by putting everything into cm:
3.12 ft = 95.10 cm
0.0455 yd = 4.07 cm
7.87 in = 19.99 cm

Now find the volume:
V = l w h
V = (95.1 cm)(4.07 cm)(19.99 cm)
V = 7737.27 cm³

= 7737.27 cm³

Problem 4.8

If the density of an object is 2.87 x 10⁻⁴ lbs/cubic inch, what is its density in g/mL?

2.87 x 10⁻⁴ ~~lb~~	kg	~~in~~	~~in~~	~~in~~	mL	1000 g
in³	2.20 ~~lb~~	2.54 ~~cm~~	2.54 ~~cm~~	2.54 ~~cm~~	~~cm³~~	kg

$$= 7.96 \times 10^{-3} \text{ g/mL}$$

Problem 4.9

You are baking French pastries according to a traditional recipe, which requires cooking at 230 degrees Celsius. What temperature must you set your oven to in the US?

We use the formula to convert from °C to °F:

T[°F] = ⁹/₅ T[°C] + 32
T[°F] = ⁹/₅ (230) + 32
T[°F] = 446

$$= 446 \text{ °F}$$

Problem 4.10

Teflon® coatings on cookware are rated to temperatures of 260 degrees Celsius. Can you safely put a Teflon® coated frying pan in the oven on broil? (note: US ovens define broil as >500 °F)

We use the formula to convert from °C to °F:

T[°F] = ⁹/₅ T[°C] + 32
T[°F] = ⁹/₅ (260) + 32
T[°F] = 500.0

The maximum temperature is 500 °F, so we cannot broil with our Teflon®

Problem 4.11

Convert a force of 10 Newtons (SI system) to dynes (CGS system).

1 Newton is a kg·m/s², and a dyne is a g·cm/s². Knowing that, this is becomes a straightforward conversion:

10 ~~kg·m~~	1000 g	100 cm
s²	~~kg~~	~~m~~

$$= 1 \times 10^6 \text{ dyne}$$

Problem 4.12
Convert a force of 10 Newtons to its equivalent in lb_f.

1 Newton is a $kg \cdot m/s^2$, and a lb_f is $32.17\ lb \cdot ft/s^2$. Knowing that, this is becomes a straightforward conversion:

$$\frac{10\ \cancel{kg \cdot m}}{s^2} \mid \frac{2.20\ lb}{\cancel{kg}} \mid \frac{3.28\ ft}{\cancel{m}} \mid \frac{lb_f}{32.17\ lb \cdot ft/s^2}$$

$$\boxed{= 2.24\ lb_f}$$

Problem 4.13
A cleaning tool can process 3 boxes in an hour. Each box contains 25 units. If a cleaning tool can work for 85% of the time, how many tools are needed for a factory to process 25,000 units each week?

First, solve for the units one tool can process:

$$\frac{.85\ \cancel{week}}{week} \mid \frac{7\ \cancel{day}}{\cancel{week}} \mid \frac{24\ \cancel{hr}}{\cancel{day}} \mid \frac{3\ \cancel{boxes}}{\cancel{hr}} \mid \frac{25\ units}{\cancel{box}}$$

= 10710 units/week per tool

Next, determine the number of tools needed:

$$\frac{25{,}000\ units/week}{10710\ \frac{units/week}{tool}}$$

= 2.33 tools → We cannot have a partial tool, so the factory needs:

$$\boxed{= 3\ tools}$$

Problem 4.14
The barrel is a measurement of volume that can be extremely confusing, as it has different definitions based on the context in which it is used. In the oil industry, a barrel is defined as 42 US gallons; in the beverage industry, a fluid barrel is defined as 31.5 US gallons; but in the UK a beer barrel is defined as 36 imperial gallons (43.2342 gallons). Using a spreadsheet, build a unit converter for these various measurements.

Equations and setup

	A	B	C	D	E	F
1	Inputs			Outputs		
2				Oil barrel	US Fluid Barrel	UK Beer Barrel
3	Oil barrel			=B3	=B3*42/31.5	=B3*42/43.2342
4	US Fluid Barrel			=B4*31.5/42	=B4	=B4*31.5/43.2342
5	UK Beer Barrel			=B5*43.2342/42	=B5*43.2342/31.5	=B5

Final table

	A	B	C	D	E	F
1	Inputs			Outputs		
2				Oil barrel	US Fluid Barrel	UK Beer Barrel
3	Oil barrel	100		100.00	133.33	97.15
4	US Fluid Barrel	100		75.00	100.00	72.86
5	UK Beer Barrel	100		102.94	137.25	100.00

Problem 4.15

You have the option of two cars. The first is a US-made car that costs $21,700 and gets 28 miles per gallon. The second is a European-made car that costs $28,400 and gets 19 kilometers per liter. With the cost of gasoline at $3.50 a gallon, how many miles would you have to drive the more fuel-efficient car to make up for the higher cost?

First convert the European car to miles/gallon:

19 km	3.85 L	mi
L	gal	1.6 km

= **45.7 miles/gallon**

Difference in mileage = **17.7 mi/gal**
Difference in price = **$6700**

$6700	~~gal~~	17.7 mi
	$3.50	~~gal~~

= **33,882 miles**

Problem 4.16

Express the quantities below in correct scientific notation.
- a. 41,500
- b. 0.0000324
- c. 842
- d. 54,835,000
- e. 0.03145

- a. 4.15×10^4
- b. 3.24×10^{-5}
- c. 8.42×10^2
- d. 5.4835×10^7
- e. 3.145×10^{-2}

Problem 4.17

Express the quantities below in correct scientific notation.
- a. 3.14
- b. 550.36
- c. 53,812.900
- d. 100.00001
- e. 0.0031

- a. 3.14
- b. 5.5036×10^2
- c. 5.38129×10^4
- d. 1.0000001×10^2
- e. 3.1×10^{-3}

Problem 4.18
Express the quantities below in standard form.
   a. 1.05×10^3
   b. 4.2175×10^8
   c. 3.14×10^{-4}
   d. 8.002×10^{-2}
   e. 5.185×10^2

   a. **1050**
   b. **421,750,000**
   c. **0.000314**
   d. **0.08002**
   e. **518.5**

Problem 4.19
Express the quantities below in standard form.
   a. 6.2×10^{-1}
   b. 4.569×10^6
   c. 1.8×10^0
   d. 6.4×10^{-5}
   e. 8.23×10^3

   a. **0.62**
   b. **4,569,000**
   c. **1.8**
   d. **0.000064**
   e. **8230**

Problem 4.20
Multiply or divide as indicated and express in correct scientific notation.
   a. $(2.21 \times 10^7)(3.10 \times 10^{-3})$
   b. $\dfrac{9.12 \times 10^{-5}}{3.8 \times 10^6}$
   c. $\dfrac{(9.12 \times 10^3)(9.12 \times 10^5)}{3.8 \times 10^6}$

   a. **6.85×10^4**
   b. **2.4×10^{-11}**
   c. **2.18×10^4**

Chapter 4 | Units and Conversions

Problem 4.21
Multiply or divide as indicated and express in correct scientific notation.

    a. $(2.21 \times 10^7)(3.10 \times 10^{-3})$

    b. $\dfrac{9.12 \times 10^{-5}}{3.8 \times 10^6}$

    c. $\dfrac{(9.12 \times 10^3)(9.12 \times 10^5)}{3.8 \times 10^6}$

a. **7.89×10^1**
b. **2.4×10^1**
c. **2.16×10^{-1}**

Problem 4.22
Express the following numbers in engineering notation.

    a. 3.145×10^{-1}
    b. 68.2×10^2
    c. 0.000 0052

a. **314.5×10^{-3}**
b. **68.2×10^2**
c. **5.2×10^{-6}**

Problem 4.23
Express the following numbers in engineering notation.

    a. 53,800
    b. 0.814
    c. 1,625,000,000

a. **53.8×10^3**
b. **814×10^{-3}**
c. **1.625×10^9**

Problem 4.24
Write the following in SI Units using the preferred prefix.

    a. 6000 g
    b. 0.005 m

a. **6 kg**
b. **5 mm**

Problem 4.25
Write the following in SI Units using the preferred prefix.
- a. 0.0003 A
- b. 0.00009625 m

    a. 300 µA
    b. 96.25 µm

Problem 4.26
Complete the following table:

Quantity	Engineering Notation	SI unit w/ prefix	Scientific Notation
510 m			
84,200 m			
0.000006 m			
58,212,000 m			
0.900 m			

Quantity	Engineering Notation	SI unit w/ prefix	Scientific Notation
510 m	510 m	510 m	5.1×10^2 m
84,200 m	84.2×10^3 m	84.2 km	8.42×10^4 m
0.000006 m	7×10^{-6} m	7 µm	7×10^{-6} m
58,212,000 m	58.212×10^6 m	58.212 Mm	5.8212×10^7 m
0.900 m	900×10^{-3}	900 mm	9.00×10^{-1} m

Problem 4.27
1.3 mg of pollutant is found in an analysis of 2500 kg of soil. What is the concentration of pollutant in parts-per notation?

We first put the measurements in like units:
1.3 mg = 1.3×10^{-3} g
2500 kg = 2.5×10^6 g

Now determine the concentration:
$$\frac{1.3 \times 10^{-3} \text{ g pollutant}}{2.5 \times 10^6 \text{ g sample}} = 5.2 \times 10^{-10} = 0.52 \times 10^{-9}$$

The exponent then determines the parts-per scale:
0.52×10^{-9} = 0.52 ppb

> Contaminant concentration is 0.52 ppb

Problem 4.28

A 2.4 kg sample of gasoline is found to have a metallic concentration of 20 ppm. What is the mass of metallic contaminants in the sample?

The exponent determines the parts-per scale:
20 ppm = 20×10^{-6}

We know the concentration in the sample:
$$\frac{\text{kg pollutant}}{2.4 \text{ kg sample}} = 20 \times 10^{-6}$$
kg pollutant = $20 \times 10^{-6} \cdot 2.4$ kg = 48×10^{-6} kg

> **The mass of metallics in the sample is 48 mg**

Problem 4.29

Given the below table, determine the number of moles of the various trace components present in 1000 kg of the earth's atmosphere.

Component	Formula	Concentration
Sulfur Dioxide	SO_2	1.0 ppm
Methane	CH_4	2.0 ppm
Nitrous Oxide	N_2O	0.5 ppm
Ozone	O_3	0.07 ppm
Nitrogen Dioxide	NO_2	0.02 ppm
Iodine	I_2	0.01 ppm

To solve this problem, we will take the following steps:
1. *Determine the actual mass of each component in 1000 kg of air*
2. *Convert that mass to grams*
3. *Divide the mass (in g) by the molecular weight to determine moles present*

Equations and setup

	A	B	C	D	E	F	G
1	Component	Formula	Concentration (ppm)	Mass in 1000 kg of air (kg)	Mass in g	Molecular Weight (g/mol)	mol in 1000 kg of air
2	Sulfur Dioxide	SO_2	1.000	=1000*C13*10^-6	=D13*1000	64.066	=E13/F13
3	Methane	CH_4	2.000	=1000*C14*10^-6	=D14*1000	16.040	=E14/F14
4	Nitrous Oxide	N_2O	0.500	=1000*C15*10^-6	=D15*1000	44.013	=E15/F15
5	Ozone	O_3	0.070	=1000*C16*10^-6	=D16*1000	48.000	=E16/F16
6	Nitrogen Dioxide	NO_2	0.020	=1000*C17*10^-6	=D17*1000	46.006	=E17/F17
7	Iodine	I_2	0.010	=1000*C18*10^-6	=D18*1000	253.809	=E18/F18

Final Table

	A	B	C	D	E	F	G
1	Component	Formula	Concentration (ppm)	Mass in 1000 kg of air (kg)	Mass in g	Molecular Weight (g/mol)	mol in 1000 kg of air
2	Sulfur Dioxide	SO_2	1.000	0.001	1.000	64.066	0.0156089
3	Methane	CH_4	2.000	0.002	2.000	16.040	0.1246883
4	Nitrous Oxide	N_2O	0.500	0.0005	0.500	44.013	0.0113603
5	Ozone	O_3	0.070	0.00007	0.070	48.000	0.0014583
6	Nitrogen Dioxide	NO_2	0.020	0.00002	0.020	46.006	0.0004347
7	Iodine	I_2	0.010	0.00001	0.010	253.809	0.0000394

Problem 4.30

A solution is prepared by mixing 50.0 g of water (H_2O) and 50.0 g of ethanol (C_2H_5OH). Determine the mole fractions of each substance.

First determine the moles of each substance:
H_2O : 25.0 g / 18.0 g/mol = 1.39 mol H_2O
C_2H_5OH : 25.0 g / 46.07 g/mol = 0.543 mol C_2H_5OH
Total mol : 1.39 mol + 0.543 mol = 1.933

Next determine the mole fractions:
H_2O : 1.39 mol / 1.933 mol = 0.719
C_2H_5OH : 0.543 mol / 1.933 mol = 0.281

$$Y_{H2O} = 0.719$$
$$Y_{C2H5OH} = 0.281$$

Problem 4.31

A solution is prepared by mixing 1 mol of water (H_2O) and 1 mol of ethanol (C_2H_5OH). Determine the mass fractions of each substance.

First determine the mass of each substance:
H_2O : 1 mol · 18.0 g/mol = 18.0 g H_2O
C_2H_5OH : 1 mol · 46.07 g/mol = 46.07 g C_2H_5OH
Total mass : 18.0 g mol + 46.07 g = 64.07 g

Next determine the mass fractions:
H_2O : 18.0 g / 64.07 g = 0.281
C_2H_5OH : 46.07 g / 64.07 g = 0.719

$$X_{H2O} = 0.281$$
$$X_{C2H5OH} = 0.719$$

Problem 4.32

Given the below table, determine the molar composition of air in the earth's atmosphere.

Component	Formula	Mass %
Oxygen	O_2	23.20
Nitrogen	N_2	75.47
Carbon Dioxide	CO_2	0.046
Argon	Ar	1.28
Neon	Ne	0.0012
Helium	He	0.00007
Krypton	Kr	0.003
Xenon	Xe	0.00004

To solve this problem, we will take the following steps:
1. Assume a basis of calculation of 1000 g of air
2. Fine the relative masses
3. Convert to relative moles
4. Determine the molar fractions

Equations and setup

	A	B	C	D	E	F	G	H
1	Basis:	1000 g						
2								
3	Component	Formula	Mass %	Mass Frac	Rel. Mass	Mol. wt (g/mol)	Rel. Moles	Mol %
4	Oxygen	O_2	23.2	=C4/100	=D4*$B$1	32.000	=E4/F4	=G4/$G$13
5	Nitrogen	N_2	75.47	=C5/100	=D5*$B$1	28.014	=E5/F5	=G5/$G$13
6	Carbon Dioxide	CO_2	0.046	=C6/100	=D6*$B$1	44.010	=E6/F6	=G6/$G$13
7	Argon	Ar	1.28	=C7/100	=D7*$B$1	39.394	=E7/F7	=G7/$G$13
8	Neon	Ne	0.0012	=C8/100	=D8*$B$1	20.198	=E8/F8	=G8/$G$13
9	Helium	He	0.00007	=C9/100	=D9*$B$1	4.003	=E9/F9	=G9/$G$13
10	Krypton	Kr	0.003	=C10/100	=D10*$B$1	83.798	=E10/F10	=G10/$G$13
11	Xenon	Xe	0.00004	=C11/100	=D11*$B$1	131.293	=E11/F11	=G11/$G$13
12								
13			Sum:	=SUM(C4:C11)	=SUM(D4:D11)		=SUM(G4:G11)	=SUM(H4:H11)

Final Table

	A	B	C	D	E	F	G	H
1	Basis:	1000 g						
2								
3	Component	Formula	Mass %	Mass Frac	Rel. Mass	Mol. wt (g/mol)	Rel. Moles	Mol %
4	Oxygen	O_2	23.2	0.232	232	32.000	7.25	0.20998299
5	Nitrogen	N_2	75.47	0.7547	754.7	28.014	26.9401	0.780270764
6	Carbon Dioxide	CO_2	0.046	0.00046	0.46	44.010	0.010452	0.000302728
7	Argon	Ar	1.28	0.0128	12.8	39.394	0.324923	0.009410788
8	Neon	Ne	0.0012	0.000012	0.012	20.198	0.000594	1.72075E-05
9	Helium	He	0.00007	0.0000007	0.0007	4.003	0.000175	5.06476E-06
10	Krypton	Kr	0.003	0.00003	0.03	83.798	0.000358	1.03689E-05
11	Xenon	Xe	0.00004	0.0000004	0.0004	131.293	3.05E-06	8.82398E-08
12								
13			Sum:	100.000	1.000		34.527	1.000

Chapter 5 Solutions
ELECTRICAL CIRCUITS

Problem 5.1
Over the course of 8 hours, 7.6×10^4 C of charge passes through a resistor. What is the current through that resistor?

An Ampere is a rate of charge - Coulombs per second. To determine the current, we divide the total charge moving through the resistor by the time that it took to do so.

A = C/s

I = 7.6×10^4 / ((8 hr) (60 min/hr) (60 sec/min))

I = 7.6×10^4 / 28800 sec

$\boxed{I = 2.63 \text{ A}}$

Problem 5.2
A load draws 20 amps for a 12 hour period. Determine the total charge that flows through the load during this period.

An Ampere is a rate of charge - Coulombs per second. To determine the total charge, we multiply the current through the load by the time that it was allowed to flow for.

C = A · time
C = 20 A * ((12 hr) (60 min/hr) (60 sec/min))
C = 20 A * 43200 sec

$\boxed{C = 8.64 \times 10^5 \text{ Coulombs}}$

Problem 5.3
What is the total voltage in a circuit if the current is 0.75 A and the resistance is 35 Ω?

This is a simple Ohm's Law problem.

V = IR
V = (0.75 A) (35 Ω)

$\boxed{V = 26.25 \text{ V}}$

Problem 5.4

How much resistance is in a circuit with 120 V and a current of 0.85 A?

This is a simple Ohm's Law problem.

V = IR
R = V / I
R = 120 V / 0.85 A

$\boxed{R = 141.18 \, \Omega}$

Problem 5.5

A blender has an 8-amp motor. What is the resistance of the blender when you plug it in to your 120-volt kitchen outlet?

This is a simple Ohm's Law problem.

V = IR
R = V / I
R = 120 V / 8 A

$\boxed{R = 15 \, \Omega}$

Problem 5.6

A coffee maker has a heating coil with 10 Ω of resistance. What is the current of the through the coil when you plug it in to your 120-volt kitchen outlet?

This is a simple Ohm's Law problem.

V = IR
I = V / R
I = 120 V / 10 Ω

$\boxed{I = 12 \, A}$

Problem 5.7

Using the below diagram, which switches must be closed in order to light up the bulb? Which switches must be closed in order to flow current but not have the bulb light up?

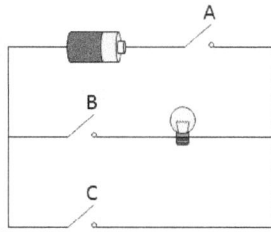

To light up the bulb, switch A and switch B must be closed to complete a circuit through the bulb. Switch C can be either open or closed; it is not relevant in this case.

To flow current through the circuit but not through the bulb, switch A and switch C must be closed, and switch B must be open.

Problem 5.8

Construct a circuit diagram that consists of a voltage source and a single resistor in series with a set of two resistors in parallel.

The circuit diagram below meets the criteria outlined in the problem.

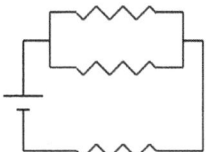

Problem 5.9

Construct a circuit diagram that consists of a voltage source and a single resistor in series with a set of two resistors in parallel.

The circuit diagram below meets the criteria outlined in the problem.

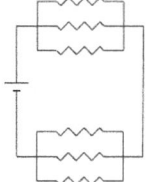

Problem 5.10

Using the below diagram, determine which resistors:
- a. Have the same current?
- b. Have the same voltage drop?

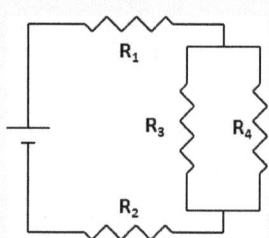

a) R_1 and R_2 are in series, and thus have the same current. *Note that the R_3/R_4 grouping also has the same current, but those individual resistors do not.*

b) R_3 and R_4 are in parallel, and thus have equivalent voltage drops.

Problem 5.11

Determine the equivalent resistance of a 4 Ω and 6 Ω resistor if they are:
- a. Connected in series
- b. Connected in parallel

In series the resistances are additive:

$R_T = R_1 + R_2$
$R_T = 4\,\Omega + 6\,\Omega$

$\boxed{R_T = 10\,\Omega}$

In parallel the inverse resistances are additive:

$1/R_T = 1/R_1 + 1/R_2$
$1/R_T = 1/4\,\Omega + 1/6\,\Omega$

$\boxed{R_T = 2.4\,\Omega}$

Problem 5.12

Two resistors of 4 Ω and 6 Ω are connected to a 12 V power source. Determine the overall current if the resistors are:
- a. Connected in series
- b. Connected in parallel

In series the resistances are additive:

$R_T = R_1 + R_2$
$R_T = 4\,\Omega + 6\,\Omega$
$R_T = 10\,\Omega$

And then it becomes a simple Ohm's Law problem:

$V = IR$
$I = V / R$
$I = 12\,V / 10\,\Omega$
$\boxed{I = 1.2\,A}$

In parallel the inverse resistances are additive:

$1/R_T = 1/R_1 + 1/R_2$
$1/R_T = 1/4\,\Omega + 1/6\,\Omega$
$R_T = 2.4\,\Omega$

And then it becomes a simple Ohm's Law problem:

$V = IR$
$I = V / R$
$I = 12\,V / 2.4\,\Omega$
$\boxed{I = 5\,A}$

Problem 5.13

The circuit below consists of three resistors connected to a power source. The voltage source is a 20 V source and the resistors are 3 Ω, 4 Ω, and 8Ω. Determine the current in the circuit, I_T.

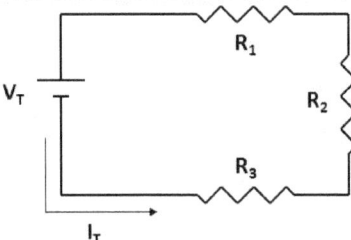

As per the diagram, all of the resistors are in series. This is this a straightforward Ohm's Law problem.

$V = IR$
$I_T = V_T / (R_1 + R_2 + R_3)$
$I_T = 20 \text{ V} / (3 \text{ Ω} + 4 \text{ Ω} + 8 \text{ Ω})$

$I_T = 1.33 \text{ A}$

Problem 5.14

The circuit shown consists of three resistors connected to a 120 V power source. The current, I_T, in the circuit is 30 A. If two of the resistors are the same resistance, and the third has twice the resistance, solve for R_1, R_2, and R_3.

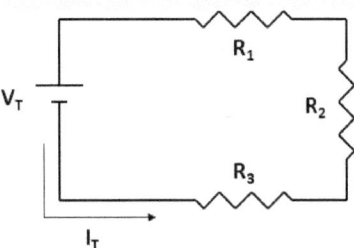

First step is to solve for the resistance in the circuit, which is a straightforward Ohm's Law problem.

$R_T = V_T / I_T$
$R_T = 120 \text{ V} / 30 \text{ A}$
$R_T = 40 \text{ Ω}$

As per the diagram, all of the resistors are in series. The rest of the solution is just algebra.

$R_T = (R_1 + R_2 + R_3)$
$40 \text{ Ω} = (X + X + 2X)$
$10 \text{ Ω} = X$

One resistor is 20 Ω and the other two are 10 Ω each.

Problem 5.15

The circuit below consists of three resistors connected to a power source. The voltage source is a 20 V source and the resistors are 3 Ω, 4 Ω, and 8Ω. Determine the current in the circuit, I_T.

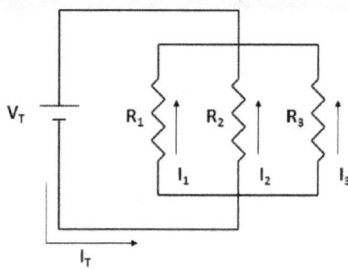

First step is to solve for the resistance in the circuit, which as per the diagram are all in parallel.

$1/R_T = 1/R_1 + 1/R_2 + 1/R_3$
$1/R_T = 1/3\ \Omega + 1/4\ \Omega + 1/8\ \Omega$
$1/R_T = 0.708$
$R_T = 1.412$

We can then solve for the current using Ohm's Law.

$V = IR$
$I_T = V_T / R_T$
$I_T = 20\ V / 1.412$

$I_T = 14.16\ A$

Problem 5.16

The circuit shown consists of three resistors connected to a 120 V power source. The total current, I_T, in the circuit is 30 A. If two of the resistors are the same resistance, and the third has twice the resistance, solve for R_1, R_2, and R_3.

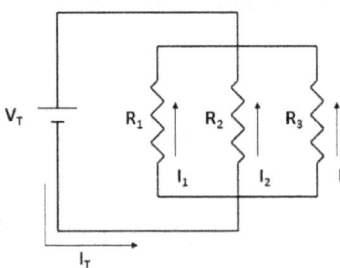

First step is to solve for the resistance in the circuit, which is a straightforward Ohm's Law problem.

$R_T = V_T / I_T$
$R_T = 120 \text{ V} / 30 \text{ A}$
$R_T = 40 \text{ }\Omega$

As per the diagram, all of the resistors are in parallel. The rest of the solution is just algebra.

$$\frac{1}{R_T} = \frac{1}{R_1} + \frac{1}{R_2} + \frac{1}{R_3}$$

$$\frac{1}{R_T} = \frac{1}{X} + \frac{1}{X} + \frac{1}{2X}$$

$$\frac{1}{R_T} = \frac{2}{2X} + \frac{2}{2X} + \frac{1}{2X}$$

$$\frac{1}{R_T} = \frac{5}{2X}$$

$$X = \frac{5}{2} R_T = \frac{5}{2}(40 \text{ }\Omega)$$

$X = 100 \text{ }\Omega$

One resistor is 200 Ω and the other two are 100 Ω each.

Problem 5.17

The circuit below consists of four resistors connected to a power source. The voltage source is a 20 V source and the series resistors are 3 Ω and 4 Ω. The resistors in parallel are 8Ω and 12Ω. Determine the current in the circuit, I_T.

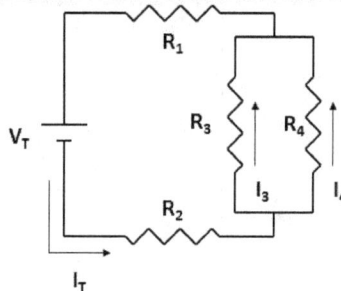

First step is to solve for the total resistance in the circuit, which as per the diagram is a combination circuit. We start by solving for the equivalent resistance of the parallel resistors.

$1/R_P = 1/R_1 + 1/R_2$
$1/R_P = 1/8\ \Omega + 1/12\ \Omega$
$1/R_P = 0.308$
$R_P = 4.8$

$R_T = (R_1 + R_2 + R_P)$
$R_T = 3\ \Omega + 4\ \Omega + 4.8\ \Omega$
$R_T = 11.8\ \Omega$

We can then solve for the current using Ohm's Law.

$V = I R$
$I_T = V_T / R_T$
$I_T = 20\ V / 11.8\ \Omega$

$I_T = 1.70\ A$

Problem 5.18

What is the power generated by the blender in problem #5 of this chapter?

To find the power, we multiply the current by the voltage.

$P = IV$
$P = (8\ A)(120\ V)$

$P = 960\ W$

Chapter 5| Electrical Circuits

Problem 5.19

What is the power generated by the heating coil in problem #6 of this chapter?

To find the power, we multiply the resistance by the square of the current.

$P = V^2 / R$
$P = (120 \text{ V})^2 / (10 \text{ Ω})$

$\boxed{P = 1440 \text{ W}}$

Problem 5.20

If it takes 30 seconds to blend your smoothie, then what is the energy consumed by the blender in problem #3 of this chapter?

To find the energy, we multiply the square of the current by the resistance and by the duration.

$Q = I^2 R t$
$Q = (0.75 \text{ A})^2 (35 \text{ Ω}) (30 \text{ s})$

$\boxed{Q = 590.6 \text{ J}}$

Problem 5.21

If it takes 5 minutes to make a cup of coffee, then what is the energy consumed by the coffee maker in problem #4 of this chapter?

Energy is power multiplied by duration (current times electricity times duration).

$Q = I V t$
$Q = (0.85 \text{ A}) (120 \text{ V}) (300 \text{ s})$

$\boxed{Q = 30.6 \times 10^3 \text{ kW}}$

Problem 5.22

Your flat screen TV consumed 345 kJ of energy over the course of a 180-minute movie. What is the power rating of the television set?

Energy is power multiplied by duration.

Q = Power (time)
Power = Q / time
Power = ((345 kJ) (1000 J/kJ)) / ((180 min) (60 s/min))

$$\boxed{P = 31.94 \text{ W}}$$

Problem 5.23

Your living room is illuminated by four 60 Watt light bulbs. If your lights are on for 6 hours each night, how much energy is consumed over the course of a week if:
    a. The lights are connected in series
    b. The lights are connected in parallel

This is a trick question. While the overall current will be different based on series/parallel configurations, because we know the power rating of each bulb the energy consumption can be directly calculated.

Q = Power (time)
Q = (4 bulbs) (60 W/bulb) (6 hrs/day) (7 days/wk) (60 min/hr) (60 sec/min)

$$\boxed{Q = 3.6288 \times 10^7 \text{ J consumed per week}}$$

Chapter 5 | Electrical Circuits

Problem 5.24

General Electric's 42-Watt compact fluorescent light bulb (CFL) advertises the equivalent illumination as a standard 150-watt incandescent bulb. However, the CFL bulb costs $10.97, whereas the incandescent bulb only costs $3.97 for a two-pack. If you pay $0.103 per kWh, how many hours of use will it take to make up the cost difference?

Cost per hour for the CFL bulb:
= (42 W) ($0.103 kW/hr) (kW/1000 W)
= **$0.004326 / hr**

Cost per hour for the incandescent (IND) bulb:
= (150 W) ($0.103 kW/hr) (kW/1000 W)
= **$0.01545 / hr**

Set the two equal to each other at duration of time X:
CFL base cost + (CFL $/hr) (hrs of use) = IND base cost + (IND $/hr) (hrs of use)
$10.97+ ($0.004326/hr) (X hr) = $1.985 + ($0.01545/hr) (X hr)
X hrs = $8.985 / ($0.011124/hr)
X = 807 hr

> You recover the cost difference of the CFL bulb after ~807 hours of use.

Problem 5.25

A single string of incandescent Xmas lights is 70 ft long and rated at 72 W. If you have 350 ft of lights, and you run your lights for 6 hours a day for 30 days during the holiday season, what is the total cost of 'holiday cheer' if you pay $0.103 per kWh?

Number of light strings:
= (350 ft) / (70 ft/string)
= **5 strings**

Total number of hours per season:
= 6 hrs/day · 30 days/season
= **180 hrs/season**

Total cost of energy:
= Cost of energy per hour · # of strings · hours per season · energy per string
= $0.103/kWh · 5 strings · 180 hr/season · 72 W/string · kW/1000 W
= **$6.67/season**

> It will cost ~ $6.67 for the holiday season

Problem 5.26

A single string of LED Xmas lights is 50 ft long and rated at 7.2 W. If you have 350 ft of lights, and you run your lights for 6 hours a day for 30 days during the holiday season, what is the total cost of 'holiday cheer' if you pay $0.103 per kWh?

Number of light strings:
= (350 ft) / (50 ft/string)
= 7 strings

Total number of hours per season:
= 6 hrs/day · 30 days/season
= 180 hrs/season

Total cost of energy:
= Cost of energy per hour · # of strings · hours per season · energy per string
= $0.103/kWh · 7 strings · 180 hr/season · 7.2 W/string · kW/1000 W
= $6.67/season

> It will cost only ~ $0.93 cents for the holiday season

Problem 5.27

If the incandescent string in problem #24 of this chapter costs $8.98, and the LED string in problem #25 of this chapter costs $22.98, what is the overall cheaper option for your 'holiday cheer'?

Total cost of the incandescent lights:

= 5 · $8.98 + $6.67
= $51.57

Total cost of the LED lights:

= 7 · $22.98 + $0.93
= $161.79

> The LEDs are more energy-efficient, but are overall the more expensive option.

Problem 5.28

A circuit breaker is placed in series with the parallel branches on a kitchen circuit, ensuring that the current on the circuit does not reach dangerous levels. In the setup below, the circuit breaker is rated for 20 A, and has a 120 V source. If your kitchen has a 21 Ω blender, a 12 Ω coffee maker, a 16 Ω toaster, and an 81 Ω microwave, what combination of these appliances can be operated simultaneously without tripping the breaker? (The breaker will 'trip' if the current in the circuit exceeds 20 A).

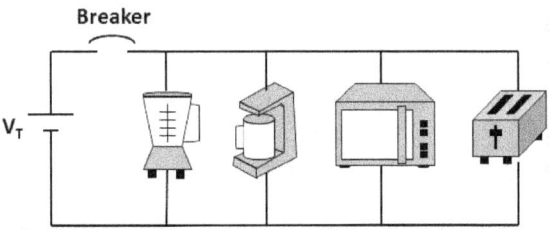

There are 10 combinations of the appliances. For a problem like this that includes multiple repetitive calculations, you may consider using a tool such as the one that was developed in the Excel learning activity.

Blender (21 Ω)	Coffee Maker (12 Ω)	Toaster (16 Ω)	Microwave (81 Ω)	R_T (Ω)	I_T (A)	Result
X				21	5.71	OK
X	X			7.64	15.7	OK
X	X	X		5.17	23.2	Fail
X	X	X	X	4.86	24.69	Fail
	X			12	10	OK
	X	X		6.86	17.5	OK
	X	X	X	6.32	18.98	OK
		X		16	7.5	OK
		X	X	13.36	8.98	OK
			X	81	1.48	OK

All combinations will work except for two:
- All four appliances at the same time
- The blender, coffee maker, and toaster at the same time

Problem 5.29

A single string of Xmas lights has a power rating of 72 W. A number of these light strings are connected in a chain (in series) and plugged into a 120-Volt outdoor receptacle. If the breaker to that outdoor receptacle trips at 20 A, what is the maximum number of light strings that circuit support?

This is essentially a simple power problem; we need to solve for the total power allowed and then solve for the number of light strings.

$P = IV$
(72 W/string) (n strings) = 20 A · 120 V
n = 33.3 strings

The circuit can support 33 strings before tripping

Problem 5.30

The diagram below shows a circuit with 10 resistors connected to a battery. Solve for:
  a. The total current in the circuit.
  b. Voltage across the 0.5 Ω resistor
  c. Current across the 0.6 Ω resistor
  d. Voltage across the 0.6 Ω resistor
  e. Current across the 3 Ω resistor

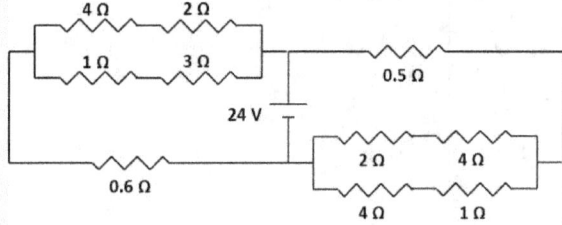

We first define our individual components. To help us keep track, we will call "A branch" will be the 0.6 Ω side, "B branch" will be the 0.5 Ω side, "resistor C" will be the parallel set of resistors on A branch, and "resistor D" will be the parallel set of resistors on B branch):

$1/R_T = 1/R_A + 1/R_B$
$R_A = 0.6\ \Omega + R_C$
$R_B = 0.5\ \Omega + R_D$
$1/R_C = 1/(2\ \Omega + 4\ \Omega) + 1/(1\ \Omega + 3\ \Omega)$
$1/R_D = 1/(2\ \Omega + 4\ \Omega) + 1/(1\ \Omega + 4\ \Omega)$

and then solve bottom-up for those components:
$R_C = 2.4\ \Omega$
$R_D = 2.22\ \Omega$
$R_A = 0.6\ \Omega + R_C = 0.6\ \Omega + 2.4\ \Omega = 3\ \Omega$
$R_B = 0.5\ \Omega + R_D = 0.5\ \Omega + 2.22\ \Omega = 2.72\ \Omega$
$1/R_T = 1/R_A + 1/R_B \rightarrow R_T = 1.28\ \Omega$

We know the voltage drop across Loop A and Loop B is 24 V, since they are in parallel. So the currents:
$I_A = V_A / R_A = 24\ V / 3\ \Omega = 8\ A$
$I_B = V_B / R_B = 24\ V / 2.72\ \Omega = 8.82\ A$

a) Total current in the circuit
$I_T = V_T / R_T = 24\ V / 1.28\ \Omega$

$\boxed{I_T = 18.75\ A}$

b) Voltage across the 0.5 Ω resistor:
$V_{0.5\Omega} = I_B\ 0.5\ \Omega = 8.82\ A \cdot 0.5\ \Omega$

$\boxed{V_{0.5\Omega} = 4.41\ V}$

c) Current across the 0.6 Ω resistor:
We have already solved for this above; the current in A branch is 8 A.

$\boxed{I_{0.6\Omega} = 8\ A}$

d) Voltage across the 0.6 Ω resistor:
$V = IR$
$V = I_{0.6\Omega}\ R = 8\ A \cdot 0.6\ \Omega$
$V = 4.8\ V$

$\boxed{V_{0.6\Omega} = 4.8\ V}$

e) Current across the 3 Ω resistor:
We know that the voltage drop across all of A branch is 24 V, and we know the voltage drop across the 0.6 Ω resistor is 4.8 A. Thus, the voltage drop across resistor C is 19.2 V.

Because the 1Ω/3Ω sub-branch is in parallel with the other sub-branch, the voltage drop across them both is 19.2 V. The current through the 1Ω/3Ω sub-branch (and thus the 3Ω resistor) is:

$I_{3\Omega} = V_{1\Omega/3\Omega} / R_{1\Omega/3\Omega}$
$I_{3\Omega} = 19.2\ \Omega / 4\ \Omega$
$I_{3\Omega} = 32\ A$

$\boxed{I_{3\Omega} = 4.8\ A}$

Problem 5.31

Given the circuit below, find the:
  a. Total current in the circuit I_T
  b. Voltage across the 4 Ω resistor
  c. Voltage across the 5 Ω resistor
  d. Current across the 2 Ω resistor
  e. Current across the 7 Ω resistor

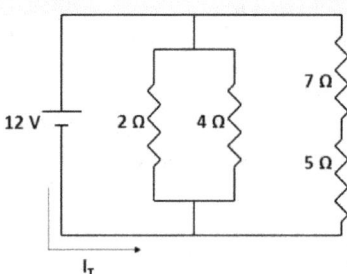

We first define our individual components. To help us keep track, we will call "A branch" 7 Ω/5 Ω branch, and "B branch" will be the 2 Ω / 4 Ω parallel resistors:
$1/R_T = 1/R_A + 1/R_B$
$R_B = 7 Ω + 5 Ω$
$1/R_A = 1/(2 Ω + 4 Ω)$

and then we solve bottom-up for those components:
$R_A = 1.33 Ω$
$R_B = 12 Ω$
$1/R_T = 1/R_A + 1/R_B$ → $R_T = 1.2 Ω$

We know the voltage drop across Loop A and Loop B is 24 V, since they are in parallel. So the currents:
$I_A = V_A / R_A = 12 V / 1.33 Ω = 9 A$
$I_B = V_B / R_B = 12 V / 12 Ω = 1 A$

a) Total current in the circuit I_T
$I_T = V_T / R_T = 12 V / 1.2 Ω$
$\boxed{I_T = 10 A}$

b) Voltage across the 4 Ω resistor:
No calculation here is needed - we know that the voltage drop across all of A branch is 12 V, and we thus know the voltage drop across the 4 Ω resistor is 12 V.
$\boxed{V_{4Ω} = 12 V}$

c) Voltage across the 5 Ω resistor:
The current in B branch is 1 A. Since the two resistors in this branch are in series, the same current passes through them. The voltage drop across this resistor is then:
$V = I R$

$V_{5\Omega} = I_B R = 1 A \cdot 5 \Omega$
$V_{5\Omega} = 5 V$

$\boxed{V_{5\Omega} = 5 V}$

d) Current across the 2 Ω resistor:

The voltage drop across A branch is 12V. Since the two resistors in this branch are in parallel, they have the same voltage across them. The current across this resistor is then:

$I_{2\Omega} = V_A / R$
$I_{2\Omega} = 12 V / 2 \Omega$
$I_{2\Omega} = 6 A$

$\boxed{I_{2\Omega} = 6 A}$

e) Current across the 7 Ω resistor:

This has already been solved. Since the two resistors in this branch are in series, the same current passes through them both.

$\boxed{I_{7\Omega} = 1 A}$

Problem 5.32

Calculate the overall resistance of 10 meters of wiring with diameter of 1 mm if it is made from:
    a. Copper
    b. Aluminum

The cross sectional area of the wire is given by:

$A_C = \pi(d/2)^2$

$A_C = (3.14)((0.5 \text{ mm})(1 \text{ m}/1000 \text{ mm}))^2$

$A_C = 7.85 \times 10^{-7} \text{ m}^2$

a) From table X.X, copper has a resistivity of $1.7 \times 10^{-8} \Omega m$

$R = \rho \dfrac{L}{A_C}$

$R = 1.7 \times 10^{-8} \Omega m \dfrac{10 \text{ m}}{7.85 \times 10^{-7} \text{m}^2}$

$\boxed{R = 0.22 \; \Omega}$

b) From table X.X, aluminum has a resistivity of $2.8 \times 10^{-8} \Omega m$

$R = \rho \dfrac{L}{A_C}$

$R = 2.8 \times 10^{-8} \Omega m \dfrac{10 \text{ m}}{7.85 \times 10^{-7} \text{m}^2}$

$\boxed{R = 0.37 \; \Omega}$

Problem 5.33

You are given specifications of copper wiring based on an electrical system's needs, but only have aluminum wire available. What size aluminum wire should you use to achieve the same system performance?

We do not need to know the specified dimensions of the copper wire to answer this problem. If we set the resistances equal to each other:

$$\rho_{Copper} \frac{L}{A_{C,Copper}} = \rho_{Aluminum} \frac{L}{A_{C,Aluminum}}$$

$$A_{C,Aluminum} = A_{C,Copper} \frac{\rho_{Aluminum}}{\rho_{Copper}}$$

$$A_{C,Aluminum} = A_{C,Copper} \frac{2.8 \times 10^{-8} \Omega m}{1.7 \times 10^{-8} \Omega m}$$

$$A_{C,Aluminum} = 1.647 \, A_{C,Copper}$$

The aluminum wire needs to have a cross sectional area that is 1.647 times that of the copper wiring. In terms of diameter:

$$\pi (d_{Aluminum}/2)^2 = 1.647 \cdot \pi (d_{Copper}/2)^2$$

$$d_{Aluminum} = \sqrt{1.647} \, d_{Copper}$$

$$d_{Aluminum} = 1.283 \, d_{Copper}$$

The aluminum wiring will need to be ~1.28 times larger in diameter than the copper wiring

Problem 5.34

Calculate the overall resistance of 10 meters of 20 gauge wiring with diameter of 1 mm if it is made from:
  a. Copper
  b. Aluminum

20 gauge wiring has a diameter of 0.8128 mm. The cross sectional area of the wire is given by:

$A_C = \pi(d/2)^2$

$A_C = (3.14) ((0.4064 \text{ mm}) (1 \text{ m}/1000 \text{ mm}))^2$

$A_C = 5.19 \times 10^{-7} \text{ m}^2$

a) From table X.X, copper has a resistivity of $1.7 \times 10^{-8} \Omega m$

$R = \rho \dfrac{L}{A_C}$

$R = 1.7 \times 10^{-8} \Omega m \dfrac{10 \text{ m}}{5.19 \times 10^{-7} \text{m}^2}$

$\boxed{R = 0.33 \, \Omega}$

b) From table X.X, aluminum has a resistivity of $2.8 \times 10^{-8} \Omega m$

$R = \rho \dfrac{L}{A_C}$

$R = 2.8 \times 10^{-8} \Omega m \dfrac{10 \text{ m}}{5.19 \times 10^{-7} \text{m}^2}$

$\boxed{R = 0.54 \, \Omega}$

Problem 5.35

Calculate the total heat generated in one minute when a 10 A current is pushed through 1 meter of 16 gauge wiring with diameter of 1 mm if it is made from:
- a. Copper
- b. Aluminum

16 gauge wiring has a diameter of 1.2903 mm. The cross sectional area of the wire is given by:
$A_C = \pi(d/2)^2$
$A_C = (3.14)((0.64515 \text{ mm})(1 \text{ m}/1000 \text{ mm}))^2$
$A_C = 1.31 \times 10^{-6} \text{ m}^2$

We can combine the equations for resistance and heat to get:
$$Q = I^2 \left(\rho \frac{L}{A_C}\right) t$$

a) *From table 5.11.1, copper has a resistivity of $1.7 \times 10^{-8} \Omega m$*

$Q = I^2 \left(\rho \frac{L}{A_C}\right) t$

$Q = 100 \text{ A}^2 \left(1.7 \times 10^{-8} \Omega m \frac{1 \text{ m}}{1.31 \times 10^{-6} \text{m}^2}\right) 60 \text{ s}$

Q = 77.86 J

b) *From table 5.11.1, aluminum has a resistivity of $2.8 \times 10^{-8} \Omega m$*

$Q = I \left(\rho \frac{L}{A_C}\right)^2 t$

$Q = 10 \text{ A} \left(2.8 \times 10^{-8} \Omega m \frac{10 \text{ m}}{1.31 \times 10^{-6} \text{m}^2}\right)^2 60 \text{ s}$

Q = 128.24 J

Problem 5.36

An incandescent light bulb is rated at 100 W for a standard 120V household source. If the tungsten filament is one-hundredth of an inch in diameter, how long is the filament?

0.01 inch = 0.000254 m

The power consumed by the light bulb is determined by the formula $P = V^2 / R$. Solving for R, we get $R = V^2 / P$.

We then plug this into the definition for resistivity, and solve for length:

$$R = \rho \frac{L}{A_C}$$

$$\frac{V^2}{P} = \rho \frac{L}{A_C}$$

$$L = A_C \frac{V^2}{\rho P}$$

$$L = \pi \left(\frac{d}{2}\right)^2 \frac{V^2}{\rho P}$$

$$L = \pi \left(\frac{0.000254 \text{ m}}{2}\right)^2 \frac{(120 \text{ V})^2}{(1.09 \times 10^{-6} \, \Omega\text{m})(100 \text{ W})}$$

$$\boxed{L = 2.13 \text{ m long}}$$

Problem 5.37

Based on your knowledge of circuits and resistivity, answer the following conceptual questions:
a) When a 100 W light bulb and a 60 W light bulb are wired in series, which one will glow brighter when the current is turned on?
b) When a 100 W light bulb and a 60 W light bulb are wired in parallel, which one will glow brighter when the current is turned on?

In all cases, the bulb which uses the most power will glow most brightly. If the two bulbs share the same current (series), then the bulb with the greater voltage drop will glow more brightly. If the two bulbs share the same voltage potential, then the bulb which draws the most current will glow more brightly.

a) **The 60 Watt bulb.** In series, they share the same current. Since the 60 W bulb has more resistance, it will produce more power than the 100 W bulb and will thus glow brighter.
b) **The 100 Watt bulb.** In parallel, they share the same voltage. As the 100 W bulb has less resistance, it will draw more current than the 60 W bulb and therefore glow brighter.

Problem 5.38

The heating coils on an electric stove are generally made from the material NiChrome 80/20, an alloy made of approximately 80% nickel and 20% chromium. You have been assigned to design a heating coil as shown below. The coil should be as cheap as possible (use the least amount of material), but can have a heat output no greater than 6 kWh to prevent cooking fire risk. If the heating coil is to be 1 meter in length, what is the minimum possible cross-section your coil can have on a 220 V stove? *Hint: use table 5.1.1.*

From table 5.1.1, the resistivity of NiChrome 80/20 is 1.09×10^{-6} Ωm.

The rate of heat generation is given by:
$$Q = \frac{V^2 t}{R}$$

and the resistance is given by:
$$R = \rho \frac{L}{A_C}$$

Combining the two equations (putting in for R):
$$Q = \frac{V^2 t}{\rho \frac{L}{A_C}}$$
$$Q = \frac{V^2 t A_C}{\rho L}$$

Now using the equation can plug in for maximum allowed heat:
$$6000 \text{ Wh} = \frac{(220 \text{ V})^2 (1 \text{ hr}) A_C}{(1.09 \times 10^{-6} \Omega \text{m})(1 \text{m})}$$
$$6000 \text{ Wh} = \frac{48{,}480 \text{ V}^2 \text{hr} \cdot A_C}{1.09 \times 10^{-6} \Omega \text{m}^2}$$
6000 Wh $= 4.440 \times 10^{10}$ Wh/m$^2 \cdot A_c$
1.35×10^{-7} m$^2 = A_c$

The maximum cross section is 0.135 mm²

Solutions Manual: Introduction to Basic Concepts in Engineering

Problem 5.39

Electrical burns are the cause of approximately 1,000 deaths per year in the US, with a large portion of these occurring in industry. If the human body comes into contact with an electrical circuit, the body can become part of that circuit. As current passes through the body, heat is generated just as it would be in any wire or load. The resistance of dry skin can be as high as 100,000 Ω which is a good insulator, but the internals of the body is only 300-1000 Ω. As current passes through the body it can generate a massive amount of heat, leading to severe burns both on the skin and internal to the body. If the average resistance of the human body is assumed to be 1000 Ω, how much current would pass through the body if a person were to come into contact with a standard 120 Volt household source? How much heat would be generated if contact were to last even for only 1 second?

This is a simple calculation using the equation for heat energy, $Q = V^2 t / R$.

$Q = V^2 t / R$
$Q = (120 \text{ V})^2 (1 \text{ s}) / 1000 \text{ Ω}$
$Q = 14.4 \text{ J}$

$\boxed{Q = 14.4 \text{ J}}$

Problem 5.40

A *high voltage* system is one that carries enough voltage to cause harm to living organisms. Repeat the calculations from problem #39 of this chapter, but this time assuming contact with a 1000 V source. How much heat is produced in this situation?

This is a simple calculation using the equation for heat energy, $Q = V^2 t / R$.

$Q = V^2 t / R$
$Q = (1500 \text{ V})^2 (1 \text{ s}) / 1000 \text{ Ω}$
$Q = 2250 \text{ J}$

$\boxed{Q = 2250 \text{ J}}$

Chapter 6 Solutions
THERMODYNAMICS

Problem 6.1
Determine the kinetic energy of a 1000 kg car that is moving at 55 mph.

Start by converting the units:
55 mph = 24.59 m/s

Kinetic energy:
KE = ½ m v²
KE = ½ (1000 kg) (24.59 m/s)²
KE = 302,334.05 J

KE = 302.33 kJ

Problem 6.2
The 1000 kg car from problem #1 of this chapter is at the top of a hill that drops 100 feet over a half-mile decline. What is its potential energy?

Start by converting the units:
100 feet = 30.48 m

Potential energy only depends on the height, not the length of the hill:
PE = m g h
PE = (1000 kg) (9.8 m/s²) (30.48 m)
PE = 298,708 J

KE = 298.70 kJ

Problem 6.3

A 747 (weight of 970,000 lbs) is flying at 500 mph at an altitude of 45,000 feet above the earth's surface. Determine the kinetic energy and the potential energy of the jet.

Start by converting all of the units to SI:
970,000 lbs = 439,985 kg
45,000 ft = 13,716 m
500 mph = 223.52 m/s

Kinetic Energy:
$KE = \frac{1}{2} m v^2$
$KE = \frac{1}{2} (439{,}985 \text{ kg}) (223.52 \text{ m/s})^2$
KE = 10,991,087,179.07 J

Potential Energy:
$PE = m g h$
$PE = (439{,}985 \text{ kg}) (9.8 \text{ m/s}^2) (13{,}716 \text{ m})$
PE = 59,141,375,748

| KE = 10,991 MJ |
| PE = 59,141 MJ |

Chapter 6 | Thermodynamics

Problem 6.4

A 1500 kg car moving at 80 km/hr drives off of a cliff into the Grand Canyon. Using a spreadsheet, determine how far the car will travel before it hits the ground 1800 m below. Generate a plot of the car's height from the ground against distance traveled from the cliff edge.

To solve this problem, we will take the following steps:
1. Enter our constants and determine a change in time
2. Calculate the car's current position from the cliff (based on equation $d = d_o + vt$)
3. Calculate the car's current downward velocity (based on equation $v = v_o + at$)
4. Calculate the car's downward distance fallen over that time period
5. Calculate the car's total distance fallen, and subtract from the initial height.

Equations and setup

	A	B	C	D	E
1	Mass of Car	1500 kg			
2	Initial speed	80 km/hr			
3	Initial speed	=B2*1000/3600	m/s		
4	dT	1 s			
5	Gravity	9.8 m/s^2			
6	Initial Height	1800 m			
7					
8					
9					
10	Time	Distance from Cliff	Distance Fallen	Downward Velocity	Height From Ground
11	0	0	0	0	=$B$6-C11
12	=A11+$B$4	=B11+$B$3*$B$4	=C11+D12*$B$4	=D11+$B$5*$B$4	=$B$6-C12

Final table

	A	B	C	D	E
1	Mass of Car	1500 kg			
2	Initial speed	80 km/hr			
3	Initial speed	22.22222222 m/s			
4	dT	1 s			
5	Gravity	9.8 m/s^2			
6	Initial Height	1800 m			
7					
8					
9					
10	Time	Distance from Cliff	Distance Fallen	Downward Velocity	Height From Ground
11	0	0	0	0	1800
12	1	22.22222222	9.8	9.8	1790.2
13	2	44.44444444	29.4	19.6	1770.6
14	3	66.66666667	58.8	29.4	1741.2
15	4	88.88888889	98	39.2	1702
16	5	111.1111111	147	49	1653
17	6	133.3333333	205.8	58.8	1594.2
18	7	155.5555556	274.4	68.6	1525.6
19	8	177.7777778	352.8	78.4	1447.2
20	9	200	441	88.2	1359
21	10	222.2222222	539	98	1261
22	11	244.4444444	646.8	107.8	1153.2
23	12	266.6666667	764.4	117.6	1035.6
24	13	288.8888889	891.8	127.4	908.2
25	14	311.1111111	1029	137.2	771
26	15	333.3333333	1176	147	624
27	16	355.5555556	1332.8	156.8	467.2
28	17	377.7777778	1499.4	166.6	300.6
29	18	400	1675.8	176.4	124.2
30	19	422.2222222	1862	186.2	-62

From our spreadsheet, we find that the car will impact with the ground at just over 18 seconds after driving off of the cliff, and would have traveled a horizontal distance of roughly 422 meters.

The final plot shows that the car picks up downward velocity over time, which is exactly what we would expect.

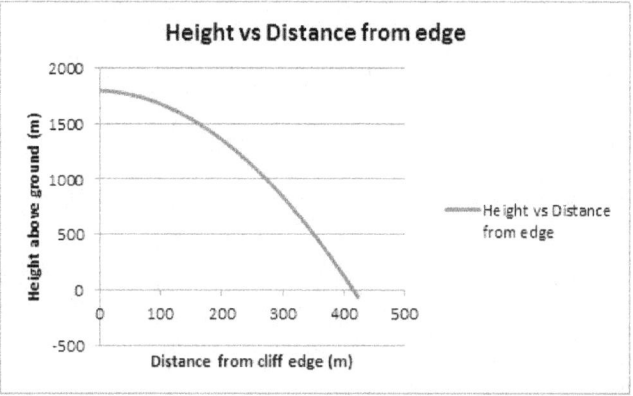

Solutions Manual: Introduction to Basic Concepts in Engineering

Problem 6.5

Consider your bedroom as a closed thermodynamic system. Your copy of "Introduction to Engineering Basics" is sitting on your desk, and your dog is burning chemical energy to wag his tail, which accidentally knocks the book to the ground. How has the overall energy of your bedroom changed?

Let's consider the components of energy change:
- As your dog wags his tail, internal energy from the system is being converted into kinetic energy (and likely some heat).
- As the book falls from the desk, potential energy is converted into heat as the book hits the ground.

However, all of this is within the thermodynamic system (the bedroom). The energy states change, but the overall total energy of the room does not.

Problem 6.6

A paddlewheel raises the temperature of a 200 mL water bath by 4 °C. How much work is done and how much heat is transferred in this process?

All of the energy transferred is done by work, thus the heat transfer is zero.

$Q = m \Delta T c$

$Q = (0.2 \text{ kg})(4 \text{ °C})(4.184 \text{ J/g °C})$

$\boxed{Q = 3.347 \text{ kJ}}$

Problem 6.7

Under the right conditions, opening an aluminum soda can will cause "spray" to appear to come out of the can. Why does this happen, and what happens to the overall energy of the soda can? Are work and heat going into or out of the can?

When the soda can is first opened, the CO_2 gas inside of the can rapidly expands, which results in a very cold region of air right outside of the can opening (think ideal gas law). This cold region only lasts for a very short period, but the moisture in the air condenses to create a 'fog'. It is not the liquid from inside the can shooting out, instead it is condensing water in the air being pushed by the rapidly expanding gas.

As the gas expands, the pop can system is effectively doing work on the surroundings. The sign of work is negative. Heat will only flow if the soda can is colder and/or warmer than the surroundings (and will have the appropriate sign).

Problem 6.8

A coiled spring releases and shoots a 0.5 kg ball 2 meters into the air. What amount of potential energy released from the coiled spring?

We first must solve for the initial kinetic energy of the ball upon release. We can do this by using a basic physics formula to solve for the initial velocity based on the final velocity:

$v_f^2 = v_i^2 + 2\,a\,d$
$v_i^2 = v_f^2 - 2\,a\,d$
$v_i^2 = 0 - 2\,(9.8\ m/s^2)\,(2\ m)$
$v_i = \sqrt{(39.2\ m^2/s^2)}$
$v_i = 6.26\ m/s$

This means that at release the ball had a KE of:
$KE = \tfrac{1}{2}\,m\,v^2$
$KE = \tfrac{1}{2}\,(0.5\ kg)\,(39.2\ m^2/s^2)$
$KE = 9.8\ J$

Due to energy conservation, we can conclude that the kinetic energy passed to the ball was due to the released energy by the spring; KE = PE.

PE = 3.347 kJ

Problem 6.9

A 3000 lb truck is moving at 55 mph at the base of a 1000 ft hill. At the top of the hill, the truck is moving 45 mph. Negating wind and road friction, what is the change in the truck's internal energy (how much energy comes from burning fuel)?

Start by converting all of the units to SI:
3000 lb = 1360.777 kg
55 mph = 24.587 m/s
45 mph = 20.116 m/s
1000 ft = 304.8 m

Any changes in kinetic and potential energy are accounted for by changes in internal energy:

$\Delta KE + \Delta PE + \Delta U = 0$
$\Delta KE + \Delta PE = -\Delta U$
$(\tfrac{1}{2}\,m v_2^2 - \tfrac{1}{2}\,m v_1^2) + (m g h_2 - m g h_1) = -\Delta U$
$\tfrac{1}{2}\,m\,(v_2^2 - v_1^2) + m g\,(h_2 - h_1) = -\Delta U$
$(\tfrac{1}{2})(1360.777\ kg)\,((20.116\ m/s)^2 - (24.587\ m/s)^2) + (1360.777\ kg)(9.8\ m/s^2)(304.8\ m) = -\Delta U$
$3{,}928{,}708.045\ J = -\Delta U$

Energy is expended to get up the hills, so since energy is leaving the system the negative sign makes sense.

ΔU = - 2928.71 kJ

Problem 6.10

A 1500 kg car moving at 55 mph on a flat road slams on its brakes and comes to a complete stop. How much heat is generated?

Start by converting all of the units to SI:
55 mph = 24.587 m/s

We first must solve for the initial kinetic energy of the car before braking.
$KE = \frac{1}{2} m v^2$
$KE = \frac{1}{2} (1500 \text{ kg})(24.587 \text{ m/s})^2$
KE = 453,390.43 J

Due to energy conservation, we can conclude that the heat energy generated is due to the initial kinetic energy of the car; KE = Q. Heat is leaving the car (to the surroundings) so it has a negative sign.

$\boxed{Q = -453.3 \text{ kJ}}$

Problem 6.11

Water at Multnomah Falls in the state of Oregon, USA, drops a vertical distance of 186 meters into the catch pool at the bottom. Consider one kilogram of water. If all of the gravitational energy that is lost during the fall is converted into thermal energy upon splashing at the bottom, how much will the water temperature increase?

First, we calculate the potential energy of the water at the top of the falls:
$PE = mgh$
$PE = (1 \text{ kg})(9.8 \text{ m/s}^2)(186 \text{ m})$
PE = 1822.8 J = 1.8228 kJ

Because of all that energy is converted to heat at the bottom of the falls, we now equate PE = Q and solve for the temperature change:
$Q = m \Delta T c$
$\Delta T = Q / (m c)$
$\Delta T = 1.8228 \text{ kJ} / [(1 \text{ kg})(4.184 \text{ kJ/kg°C})]$

$\boxed{\Delta T = 0.436 \text{ °C}}$

Problem 6.12

A 2 kg steel block at 10 °C is placed into thermal contact with a 1 kg steel block at 40 °C. Which way direction heat flow, and which sample contains more energy at final thermal equilibrium?

Heat will flow from the hot sample to the colder sample, so it will flow towards the 2 kg block. At final thermal equilibrium the larger block will contain more energy.

Problem 6.13

1 kg of steam at 100 °C is placed into thermal contact with 1 kg of liquid water at 100 °C. Which way direction heat flow, and which sample contains more energy at final thermal equilibrium?

Both samples are at the same temperature, so heat will not flow between the two. 1 kg of steam will contain more energy than a liquid sample of the same mass.

Problem 6.14

Calculate the work associated with the expansion of a gas from 25 L to 30 L against a constant external pressure of 5 atm.

We can solve first using the equation for PV work:

$$W = -P \Delta V$$
$$W = -(5 \text{ atm})(30 \text{ L} - 25 \text{ L})$$
$$W = -(5 \text{ atm})(5 \text{ L})$$
$$W = -35 \text{ atm} \cdot \text{L}$$

We then put the answer into SI units for energy:

$$\frac{-25 \text{ atm·L}}{1} \times \frac{m^3}{1000 \text{ L}} \times \frac{101{,}325 \text{ Pa}}{\text{atm}} \times \frac{N/m^2}{\text{Pa}} \times \frac{J}{N \cdot m} = -2533.125 \text{ J}$$

The gas is expanding, so we expect the work to be negative.

$$\boxed{W = -2.53 \text{ kJ}}$$

Problem 6.15

A balloon is being inflated against a constant external pressure of 1.0 atm. The balloon volume changes from 200 L to 250 L when 130 kJ of heat is added. What is the change in internal energy of the system?

The change in internal energy is given by:
$\Delta U = \Delta W + \Delta Q$

ΔQ is given in the problem, and is 130 kJ = 13,000 J.

ΔW is due to the expansion of the balloon:

$$W = -P\Delta V$$
$$W = -(1.0 \text{ atm})(250 \text{ L} - 200 \text{ L})$$
$$W = -(1.0 \text{ atm})(50 \text{ L})$$
$$W = -50 \text{ atm·L}$$

To do the addition we must put the work into SI units for energy:

$$\frac{-50 \text{ atm·L}}{} \left| \frac{m^3}{1000 \text{ L}} \right| \frac{101{,}325 \text{ Pa}}{\text{atm}} \left| \frac{N/m^2}{Pa} \right| \frac{J}{N\cdot m} = -5066.25 \text{ J}$$

The system is expanding, so the negative sign makes sense.

$\Delta U = \Delta W + \Delta Q$
$\Delta U = -5066.26 \text{ J} + 13{,}000 \text{ J}$
$\Delta U = 7933.75 \text{ J}$

The heat added is much greater than the expansion work, so the internal energy of the system increases.

$\boxed{\Delta U = 7.93 \text{ kJ}}$

Problem 6.16

100 Watts are being applied to a rotating shaft and the resulting torque is 50 Nm. How fast is the shaft spinning? Give answer in rpm.

In this case, we have a power supply rate and a torque, so we modify the equation for shaft work accordingly:

$$\dot{W}_S = \tau\, 2\pi\, \dot{n}$$

Solving for number of rotations/time:

$$\frac{\dot{W}_S}{\tau\, 2\pi} = \dot{n}$$

$$\frac{100 \text{ J/s}}{(50 \text{ Nm})\, 2\pi} = \dot{n}$$

$$\dot{n} = 0.318 \text{ rps}$$

$\boxed{\dot{n} = 19.10 \text{ rpm}}$

Problem 6.17

A particular mixer has a power rating of 1200 Watts, and at maximum speed the paddle shaft spins at 800 rpm. Calculate the torque on the shaft at the maximum setting, assuming that all power consumed goes into spinning the mixer paddle.

In this case, we have a power supply rate and a rate of rotation, so we modify the equation for shaft work accordingly:

$$\dot{W}_S = \tau \, 2 \pi \, \dot{n}$$

Solving for torque:

$$\tau = \dot{W}_S \, 2 \pi \, \dot{n}$$

$$\tau = (1200 \text{ J/s}) \, 2 \pi \, (800 \text{ rotations/\sout{min}}) \, (\sout{min}/60 \text{ s})$$

$$\tau = 6{,}031{,}857{,}89 \text{ Nm}$$

$$\boxed{\tau = 6.03 \times 10^6 \text{ Nm}}$$

Problem 6.18

Three samples of liquid are mixed together in a beaker. The temperatures of the three samples are 25 °C, 50 °C, and 75 °C. The mass of the 50 °C sample is twice the mass of the 25 °C sample, and the mass of the 75 °C sample is half of the mass of the 50 °C sample. What is the final equilibrium temperature?

Let's start by defining our samples:
Sample 1: 25 °C and mass m
Sample 2: 50 °C and mass 2m
Sample 3: 75 °C and mass m

Mixing sample 1 and sample 3:
$m \, (T_f - 25 \text{ °C}) = m \, (75 \text{ °C} - T_f)$
$2 \, T_f = 100 \text{ °C}$
$T_f = 50 \text{ °C}$

The temperature of the 1 and 3 mixture is the same as that of sample 2. Thus, the final temperature is 50 °C.

$$\boxed{T_f = 50 \text{ °C}}$$

Problem 6.19

Determine the final temperature when a 25.0 g piece of iron at 85 °C is put into 75.0 g of water at 20 °C. Assume no heat is lost to the surroundings, the heat capacity of water is constant at 4.184 J/g°C, and the heat capacity of iron is constant at 0.45 J/g°C.

The heat lost by the metal as it cools is equivalent to the heat gained by the water as it warms up:

$$Q_{lost} = Q_{gained}$$
$$M_{iron}(T_{I,iron} - T_F)c_{iron} = M_{water}(T_{I,water} - T_F)c_{water}$$

As the entire system is the same final temperature, we simply solve for T_F:

$$(25.0 \text{ g})(85 \text{ °C} - T_F)(0.45 \text{ J/g°C}) = (75.0 \text{ g})(T_F - 20 \text{ °C})(4.184 \text{ J/g°C})$$
$$956.25 \text{ J} - 11.25 \, T_F \text{ J/°C} = 313.8 \, T_F \text{ J/°C} - 6276 \text{ J}$$
$$7232.25 \text{ J} = 325.05 \, T_F \text{ J/°C}$$
$$22.25 \text{ °C} = T_F$$

$$\boxed{T_F = 22.25 \text{ °C}}$$

Problem 6.20

A 50.0 g chunk of aluminum at 300 °C is placed into 200 g of ethyl alcohol at 10 °C. Before reaching thermal equilibrium, the aluminum is removed and is found to be at 120 °C. What is the new temperature of the ethyl alcohol?

From table 6.7.1, the specific heat capacity of aluminum is 0.897 J/g °C and that of ethanol is 2.44 J/g °C.

The heat lost by the metal as it cools is equivalent to the heat gained by the alcohol as it warms up:

$$Q_{lost} = Q_{gained}$$
$$m_{aluminum}(T_{I,aluminum} - T_{F,alulminum})c_{aluminum} = m_{alcohol}(T_{F,alcohol} - T_{I,alcohol})c_{alcohol}$$

We simply solve for $T_{F,alcohol}$:

$$(50.0 \text{ g})(300 \text{ °C} - 120 \text{ °C})(0.897 \text{ J/g°C}) = (200.0 \text{ g})(T_{F,alcohol} - 10 \text{ °C})(2.44 \text{ J/g°C})$$
$$5382 \text{ J} = 488.0 \, T_{F,alcohol} \text{ J/°C} - 4880 \text{ J}$$
$$10262 \text{ J} = 488.0 \, T_{F,alcohol} \text{ J/°C}$$
$$21.03 \text{ °C} = T_{F,alcohol}$$

$$\boxed{T_{F,alcohol} = 21.03 \text{ °C}}$$

Problem 6.21

An insulated mixing tank contains 500 kg of water initially at 25 °C. If the tank is stirred for 60 minutes by a rotor spinning at 100 rpm with a torque of 1200 Nm, what is the new temperature of the liquid? Disregard the tank walls and assume no heat loss to the surroundings.

The energy generated by the spinning shaft over the 60 minute period is:

$W_S = \tau \, 2 \pi \, n$
$W_S = (1200 \text{ Nm}) \, 2\pi \, (100 \text{ rot/min}) \, (60 \text{ min})$
$W_S = 45,238.9$ kJ

All of the shaft work energy goes into raising the water temperature:

$Q = m \, \Delta T \, c$
$\Delta T = Q / (m \, c)$
$\Delta T = 45,238.9 \text{ kJ} / [(500 \text{ kg})(4.184 \text{ kJ/kg°C})]$
$\Delta T = 21.62$ °C

$\Delta T = T_F - T_I$
$\Delta T = T_F - 25$ °C

$$\boxed{T_F = 46.62 \text{ °C}}$$

Problem 6.22

What is the heat in Joules required to melt 25 kg of ice? What is the heat in calories?

From table 6.8.1 the heat of fusion of water is 334 kJ/kg.

$Q = m \cdot L_f$
$Q = 25 \text{ kg} \cdot 334 \text{ kJ/kg}$
$Q = 8350$ kJ
$Q = 8.35 \times 10^6$ J

Converting to calories:
$Q = 8.35 \times 10^6 \text{ J} \, (0.23901 \text{ cal/J})$
$Q = 1,995,733.5$ cal

$$\boxed{\begin{array}{l} Q = 8.35 \times 10^6 \text{ J} \\ Q = 2.00 \times 10^6 \text{ cal} \end{array}}$$

Problem 6.23

During a marathon, a particular runner generates heat at a rate of 400 W. Sweat evaporates from the runner's skin to help cool the body. Over the 3-hour marathon, how much water will the runner lose from sweating alone?

From table 6.8.1 the heat of vaporization of water is 2260 kJ/kg.

Over the 3 hour period at 400 Watts, the runner generates total heat energy of:
E_T = 3 hr · 400 J/s · (3600 s/hr)
E_T = 4.32 x 10^6 J

This is the energy that is available to vaporize water. Now we solve for the amount of water evaporated:
$Q = m \cdot L_v$
$m = Q / L_v$
$m = (4.32 \times 10^6 \text{ J}) / (2260 \text{ J/g})$
m = 1911.5 g

$$\boxed{m = 1.92 \text{ kg } H_2O \text{ vaporized}}$$

Problem 6.24

How much heat energy is required to convert 2 kg of ice at -40 °C into steam at 120 °C?

We first raise the temperature of the ice to 0 °C:
$Q = m \Delta T c$
$Q = (2 \text{ kg}) (40 \text{ °C}) (2.010 \text{ kJ/kg °C})$
Q = 160.8 kJ

We then melt the ice:
$Q = m L_f$
$Q = (2 \text{ kg}) (334 \text{ kJ/kg})$
Q = 668 kJ

We then raise the temperature of the water to 100 °C:
$Q = m \Delta T c$
$Q = (2 \text{ kg}) (40 \text{ °C}) (4.184 \text{ kJ/kg °C})$
Q = 836.8 kJ

We then vaporize the water:
$Q = m L_v$
$Q = (2 \text{ kg}) (2260 \text{ kJ/kg})$
Q = 4520 kJ

We then raise the temperature of the steam to 120 °C:
$Q = m \Delta T c$
$Q = (2 \text{ kg}) (20 \text{ °C}) (2.100 \text{ kJ/kg °C})$
Q = 84 kJ

Last we sum the heat due to the various segments:
$Q_{total} = Q_{solid} + Q_{melting} + Q_{liquid} + Q_{vaporization} + Q_{gas}$
$Q_{total} = 160.8 \text{ kJ} + 668 \text{ kJ} + 836.8 \text{ kJ} + 4520 \text{ kJ} + 84 \text{ kJ}$
$Q_{total} = 6269.6 \text{ kJ}$

$$\boxed{Q_{total} = 6269.6 \text{ kJ}}$$

Problem 6.25

A 3.0 g hailstone is formed in a cloud at an elevation of 2 km. Assuming all potential energy is converted to heat during the fall, what is the mass of the hailstone when it reaches the ground?

From table 6.8.1 the heat of fusion of water is 334 kJ/kg.

The potential energy of the hailstone is originally:
PE = m g h
PE = (0.003 kg) (9.8 m/s²) (2000 m)
PE = 58.8 J

We now solve for the mass of melted ice due to the energy conversion during the fall:
$Q = m \cdot L_v$
$m = Q / L_v$
m = (58.8 J) / (334 kJ/kg) · (1000 J/kJ)
m = 1.76 x 10^{-4} kg

We now solve for the mass of melted ice:
m_f = mass initial – mass melted
m_f = 3.0 g – 0.176 g
m_f = 2.824 g

$$m_{stone} = 2.824 \text{ g at the ground}$$

Problem 6.26

100 kg of steam at 150 °C is fed to a turbine, which extracts 10,000 kJ. At what temperature does the steam leave the turbine?

The heat capacity of steam is 2.1 kJ/kg °C.

The work leaving the system is equal to the heat lost by the steam. Since energy is leaving the system, the work has a negative sign.

$$
\begin{aligned}
W &= \Delta Q \\
W &= m \, \Delta T \, c \\
-10{,}000 \text{ kJ} &= 100 \text{ kg} \cdot (T_F - 150\,°C) \cdot 2.1 \text{ kJ/kg °C} \\
-47.61\,°C &= T_F - 150\,°C \\
102.39\,°C &= T_F
\end{aligned}
$$

Energy is leaving the system, so the negative sign is accurate

$$T_F = 102.39\,°C$$

Problem 6.27

100 kg of steam at 125 °C is fed to a turbine, which extracts 10,000 kJ. What is the mass fraction of water in the outlet stream?

The heat capacity of steam is 2.1 kJ/kg °C. The heat of vaporization of water is 2260 kJ/kg.

The work leaving the system is equal to the heat lost by the steam. In this case, some of that energy loss also is accounted for by condensation of steam into liquid water.

First, we must find the heat loss due to dropping the temperature of the steam to 100°C:
Q = m_{total} ΔT c
Q = 100 kg (100 °C – 125 °C) · 2.1 kJ/kg °C
Q = -5250 kJ

The remaining energy extracted is accounted for by condensation of steam into liquid water:

W	=	Q_{temp} - Q_{vap}
- 10,000 kJ	=	-5250 kJ - Q_{vap}
- 4750 kJ	=	- Q_{vap}
4750 kJ	=	m_{water} · L_v
4750 kJ	=	m_{water} · 2260 kJ/kg
2.101 kg	=	m_{water}

2.1 kg of water were condensed, so the mass fraction of water in the outlet stream is:
X_{water} = 2.1 kg / 100 kg = 0.021

X_{water} = 0.021

Problem 6.28

Steam enters a heat engine at 150 °C and leaves at 110 °C. What is the maximum efficiency of the engine?

For this equation, the temperatures must be in Kelvin.

$$\% \text{ efficiency}_{max} = \left(1 - \frac{T_{Cold}}{T_{Hot}}\right) \cdot 100$$

$$\% \text{ efficiency}_{max} = \left(1 - \frac{110 + 273}{150 + 273}\right) \cdot 100$$

$$\% \text{ efficiency}_{max} = \left(1 - \frac{383}{423}\right) \cdot 100$$

$$\% \text{ efficiency}_{max} = (1 - 0.905) \cdot 100$$

% efficiency$_{max}$ = 10.5

Problem 6.29

An electrical motor on an elevator has an efficiency of 30%. If the combined mass of the elevator and occupants is 480 kg, what is the electrical power consumption if the elevator moves upwards by 45 meters in 12 seconds?

First, we find the energy consumption required to move the elevator:
PE = m g h
PE = 480 kg · 45 m · 9.8 m/s²
PE = 211,680 J

Because it takes 12 seconds to move that distance, the power consumption is:
Power = 211,680 J / 12 second
Power = 17640 W

Now we account for the efficiency:
Total Power = 17640 W / 0.30
Total Power = 58800 W

Power = 58.8 kW

Problem 6.30

The inner and outer surface temperatures of a 5mm thick glass window are 25 and 15 °C. What is the rate of heat loss through a window that is 1m by 3m, if the thermal conductivity of glass is 0.8 W/mK.

This is a basic thermal conductivity problem.

$$\dot{Q} = \frac{k \, A \, \Delta T}{d}$$

$$\dot{Q} = \frac{(0.8 \, \frac{W}{mK})(3 m^2)(10 K)}{0.005 \, m}$$

$$\dot{Q} = 4800 \text{ W}$$

$\dot{Q} = 4800$ W

Problem 6.31

The rate of heat transfer through a wood slab 50 mm thick and with cross-section of 1 m², whose inner and outer surface temperatures are 40 and 20°C, respectively, has been determined to be 40 W. What is the thermal conductivity of the wood?

This is a basic thermal conductivity problem, but we must solve for k:

$$\dot{Q} = \frac{k \, A \, \Delta T}{d}$$

$$k = \frac{\dot{Q} \, d}{A \, \Delta T}$$

$$k = \frac{(40 W)(0.05 \, m)}{(1 m^2)(10 \, K)}$$

$k = 0.2$ W/mK

Problem 6.32

A stainless steel pipe carries water at 40 °C through a 25 °C room. Use Appendix D.3 to determine the wall thickness of a Schedule 40 pipe of NPS size 3, and calculate the rate of heat transfer through 1 meter of this pipe.

This is a basic thermal conductivity problem on a round surface, but we must use the appendices to determine the required values. From Appendix D.3, NPS3 pipe has an exterior diameter of 3.5 inches and a wall thickness on a corresponding schedule 40 pipe is 0.216 inches. From Appendix C.6, stainless steel has an approximate thermal conductivity of 16 W/mK. So:

k = 16 W/mK
r_2 = 3.5 inches = 0.0889 m
r_1 = 3.5 inches − 0.216 inches = 0.083 4136 m

$$\dot{Q} = \frac{k\, 2\pi L\, \Delta T}{\ln(r_2/r_1)}$$

$$\dot{Q} = \frac{\left(16 \frac{W}{mK}\right) 2\pi\,(1\,m)\,(15\,K)}{\ln(0.0889m / 0.0834136m)}$$

$$\dot{Q} = \frac{1507.96\,W}{\ln(1.06577)} = \frac{1507.96\,W}{0.0637}$$

$$\dot{Q} = 23672.8\,W$$

$$\boxed{\dot{Q} = 23.672\,kW}$$

Problem 6.33

An aluminum pipe carries water at 60 °C through a 25 °C room. Use Appendix D.3 to determine the wall thickness of a Schedule 80 pipe of NPS size 2, and calculate the rate of heat transfer through 1 meter of this pipe.

This is a basic thermal conductivity problem on a round surface, but we must use the appendices to determine the required values. From Appendix D.3, NPS2 pipe has an exterior diameter of 2.375 inches and a wall thickness on a corresponding schedule 40 pipe is 0.218 inches. From Appendix C.6, aluminum has a thermal conductivity of 236 W/mK. So:

k = 237 W/mK
r_2 = 2.375 inches = 0.060 325 m
r_1 = 2.375 inches − 0.218 inches = 0.054 7848 m

$$\dot{Q} = \frac{k\, 2\pi L\, \Delta T}{\ln(r_2/r_1)}$$

$$\dot{Q} = \frac{\left(237 \frac{W}{mK}\right) 2\pi\,(1\,m)\,(35\,K)}{\ln(0.060325m / 0.0547848m)}$$

$$\dot{Q} = \frac{52119.02\,W}{\ln(1.1545)} = \frac{52119.02\,W}{0.1437}$$

$$\dot{Q} = 362{,}693.26\,W$$

$$\boxed{\dot{Q} = 362.693\,kW}$$

Problem 6.34

Derive an expression for the rate of heat transfer through the slab below. Assume one-dimensional flow of heat from T_1 to T_2.

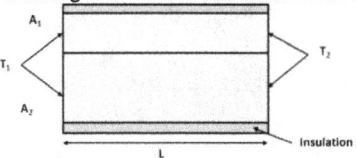

In this case, heat moves from T_1 to T_2 through the two mediums in parallel, and thus we can model this system as resistors in parallel by using the R-values of each material.

We start with the equation for rate of heat transfer:

$$\dot{Q} = \frac{k A \Delta T}{L} = \frac{\Delta T}{R_{Total}}$$

where:
$$\frac{1}{R_{Total}} = \frac{1}{R_1} + \frac{1}{R_2}$$

and:
$$R_1 = \frac{L}{k_1 A_1}$$

and:
$$R_2 = \frac{L}{k_2 A_2}$$

combining and solving for R_{Total}:

$$\frac{1}{R_{Total}} = \frac{1}{\frac{L}{k_1 A_1}} + \frac{1}{\frac{L}{k_2 A_2}} = \frac{R_1 + R_2}{R_1 R_2} = \frac{\frac{L}{k_1 A_1} + \frac{L}{k_2 A_2}}{\left(\frac{L}{k_1 A_1}\right)\left(\frac{L}{k_2 A_2}\right)} = \frac{L\left(\frac{1}{k_1 A_1} + \frac{1}{k_2 A_2}\right)}{L^2 \left(\frac{1}{k_1 A_1 k_2 A_2}\right)}$$

$$\frac{1}{R_{Total}} = \frac{1}{L} \frac{\left(\frac{k_1 A_1 + k_2 A_2}{k_1 A_1 k_2 A_2}\right)}{\left(\frac{1}{k_1 A_1 k_2 A_2}\right)} = \frac{k_1 A_1 + k_2 A_2}{L}$$

$$R_{Total} = \frac{L}{k_1 A_1 + k_2 A_2}$$

Plugging back into the main equation:

$$\dot{Q} = \frac{\Delta T}{R_{Total}} = \frac{\Delta T}{\frac{L}{k_1 A_1 + k_2 A_2}} = \frac{\Delta T}{L}(k_1 A_1 + k_2 A_2)$$

$$\boxed{\dot{Q} = \frac{\Delta T}{L}(k_1 A_1 + k_2 A_2)}$$

Problem 6.35

Derive an expression for the rate of heat transfer through the slab below. Assume one-dimensional flow of heat from T_1 to T_2. Hint: use the results of problem 34 of this chapter as a starting point.

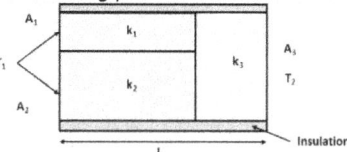

In this case, heat moves from T_1 to T_2 through the one medium in series with two mediums in parallel, and thus we can model this system as resistors in series (where one of those is the equivalent resistance of two resistors in parallel) by using the R-values of each material.

We start with the equation for rate of heat transfer:
$$\dot{Q} = \frac{k\,A\,\Delta T}{L} = \frac{\Delta T}{R_{Total}}$$

where:
$$R_{Total} = R_{12} + R_3$$

if:
$$\frac{1}{R_{12}} = \frac{1}{R_1} + \frac{1}{R_2}$$

and:
$$R_1 = \frac{L}{k_1 A_1}$$

and:
$$R_2 = \frac{L}{k_2 A_2}$$

and:
$$R_3 = \frac{L}{k_3 A_3}$$

Using the results from 6.34, we know that R_{12} is:
$$R_{12} = \frac{L}{k_1 A_1 + k_2 A_2}$$

So we can solve for R_{Total} as:
$$R_{Total} = R_{12} + R_3 = \frac{L}{k_1 A_1 + k_2 A_2} + \frac{L}{k_3 A_3}$$

$$R_{Total} = L\left(\frac{k_1 A_1 + k_2 A_2 + k_3 A_3}{k_1 A_1 k_3 A_3 + k_2 A_2 k_3 A_3}\right)$$

Plugging back into the main equation:
$$\dot{Q} = \frac{\Delta T}{R_{Total}} = \frac{\Delta T}{L\left(\frac{k_1 A_1 + k_2 A_2 + k_3 A_3}{k_1 A_1 k_3 A_3 + k_2 A_2 k_3 A_3}\right)} = \frac{\Delta T}{L}\left(\frac{k_1 A_1 k_3 A_3 + k_2 A_2 k_3 A_3}{k_1 A_1 + k_2 A_2 + k_3 A_3}\right)$$

$$\boxed{\dot{Q} = \frac{\Delta T}{L}\left(\frac{k_1 A_1 k_3 A_3 + k_2 A_2 k_3 A_3}{k_1 A_1 + k_2 A_2 + k_3 A_3}\right)}$$

Problem 6.36

Derive an expression for the rate of heat transfer through the insulated pipe shown below. Assume the interior edge of the pipe to be T_1, and one-dimensional heat transfer to T_2.

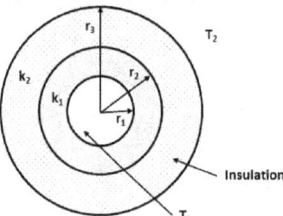

In this case, heat moves from T_1 to T_2 through the two mediums in series, and thus we can model this system as resistors in series by using the R-values of each material.

We start with the equation for rate of heat transfer for a curved surface:

$$\dot{Q} = \frac{k\, 2\pi L\, \Delta T}{\ln(r_2/r_1)} = \frac{\Delta T}{R_{Total}}$$

where: $\quad$ and: $\quad$ and:

$$R_{Total} = R_1 + R_2 \qquad R_1 = \frac{\ln(r_2/r_1)}{k_1\, 2\pi L} \qquad R_1 = \frac{\ln(r_3/r_2)}{k_2\, 2\pi L}$$

combining and solving for R_{Total}:

$$R_{Total} = R_1 + R_2 = \frac{\ln(r_2/r_1)}{k_1\, 2\pi L} + \frac{\ln(r_3/r_2)}{k_2\, 2\pi L}$$

$$R_{Total} = \frac{1}{2\pi L}\left(\frac{\ln(r_2/r_1)}{k_1} + \frac{\ln(r_3/r_2)}{k_2}\right)$$

Plugging back into the main equation:

$$\dot{Q} = \frac{\Delta T}{R_{Total}} = \frac{\Delta T}{\frac{1}{2\pi L}\left(\frac{\ln(r_2/r_1)}{k_1} + \frac{\ln(r_3/r_2)}{k_2}\right)} = \frac{2\pi L \Delta T}{\left(\frac{k_1}{\ln(r_2/r_1)} + \frac{k_1}{\ln(r_3/r_2)}\right)}$$

$$\boxed{\dot{Q} = \frac{2\pi L \Delta T}{\left(\frac{k_1}{\ln(r_2/r_1)} + \frac{k_1}{\ln(r_3/r_2)}\right)}}$$

Problem 6.37

Some window manufacturers offer triple-paned windows, touting the increased energy efficiency over double-paned windows. However, the small gain in thermal efficiency is rarely offset by the increased cost of the window. Based on your knowledge of heat transfer and thermal conductivity, why are triple-paned windows only marginally more efficient that double-paned windows, if at all?

The thermal conductivity of glass is relatively high at around 0.8 W/m°C, but air is over 30 times less at 0.024 W/m°C. Thus, the major 'insulator' in a window is the air that resides within the gap, not the glass itself. Therefore, inserting a third pane of glass will contribute only a small amount of increased thermal efficiency as compared to the overall window.

In reality, the only real way to greatly increase the thermal efficiency of a window is to increase its thickness. This gain from thickness is due to the increased amount of air, and is not due to the additional glass pane.

Dual- and triple-paned windows that are the same total thickness will have approximately the same thermal conductivity. This will hold true for four, five, and hundred-paned windows (if there were such things). The additional cost of the more complex window construction is rarely worth the energy savings.

Chapter 7 Solutions
FLUID MECHANICS

Problem 7.1
A regulation MLB baseball has a circumference of approximately 9.25 inches, and weighs approximately 5.25 ounces. Determine the density of a regulation baseball in g/cm³.

First, we find the diameter of the ball:
$C = \pi d \rightarrow d = 2.944$ in

Next we find the density:
$$\rho = \frac{mass}{volume} = \frac{mass}{\frac{4}{3}\pi\left(\frac{d}{2}\right)^3}$$

$$\rho = \frac{5.25 \text{ oz}}{\frac{4}{3}\pi\left(\frac{2.944 \text{ in}}{2}\right)^3} = \frac{5.25 \text{ oz}}{13.36 \text{ in}^3}$$

ρ = 0.393 oz/in³

Then convert into g/cm³ units:

0.393 oz	0.061 in³	28.35 g	=	**0.680 g/cm³**
in³	cm³	oz		

Problem 7.2
Under PGA regulations, a golf ball must weigh no more than 45.93 grams and have a diameter not less than 42.67 mm. Determine the density in g/cm³ of a golf ball that has the maximum allowed weight and is the minimum allowed size.

First we find the density:
$$\rho = \frac{mass}{volume} = \frac{mass}{\frac{4}{3}\pi\left(\frac{d}{2}\right)^3}$$

$$\rho = \frac{45.93 \text{ g}}{\frac{4}{3}\pi\left(\frac{42.57 \text{ mm}}{2}\right)^3} = \frac{45.93 \text{ g}}{40{,}393.32 \text{ mm}^3}$$

ρ = 0.00114 g/in³

Then convert into g/cm³ units:

0.00114 g	1000 mm³	=	**1.14 g/cm³**
mm³	cm³		

Problem 7.3

You have three identical marbles that you know each weigh 50 grams. To determine their density, you place them in a water-filled beaker with interior diameter of 5cm. The height of the water changes from 3 cm to 5 cm. What is the density a marble?

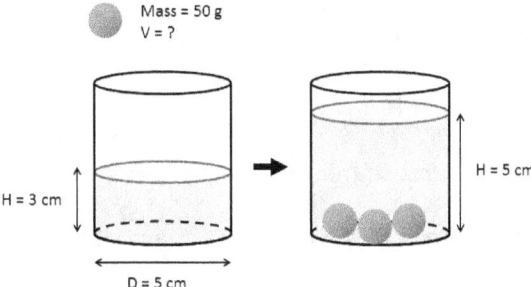

The volume change of the water is:
V = h · A = h · πr²
V = 3 cm · π(2.5 cm)²
V = 58.9 cm³

So, the change per ball is 58.9 cm³ divided by three, or 19.6 cm³. The density of a marble is then simply:

$$\rho = \frac{mass}{volume} = \frac{50 \text{ g}}{19.6 \text{ cm}^3}$$
ρ = 2.55 g/cm³

$$\boxed{\rho = 2.55 \text{ g/cm}^3}$$

Problem 7.4

Liquid A with specific gravity of 0.70 is mixed with liquid B which has specific gravity of 0.82. The final mixture has a specific gravity of 0.78. Assuming $V_{blend} = V_A + V_B$, calculate the mass ratio of A to B in the mixture.

We start by assuming a basis of calculation of 100 g of total mass. The variables then become:
$m_{mixture}$ = 100 g $V_{mixture}$ = 100 g / 0.78 g/cm³ = 128.2 cm³
m_A = 100 x_A V_A = (100 g)(x_A) / 0.70 g/cm³ = 142.9x_A
m_B = 100 x_B V_B = (100 g)(x_B) / 0.70 g/cm³ = 122.0x_B

And, because this is a two component system, we know that: 1 = x_A + x_B. → **x_A = 1 - x_B**

The equation for total volume is thus used to solve for the mass fraction of one of the components:
$V_{mixture} = V_A + V_B$
128.2 = 142.9x_A + 122.0x_B
128.2 = 142.9(1 – x_B) + 122.0x_B
-14.6 = -20.8x_B
x_B = 0.70

Solving for the other component:
x_A = 1 - x_B
x_A = 1 – 0.70
x_A = 0.30

Mass ratio A/B:
x_A/x_B = 0.30/0.70
x_A/x_B = 0.429

$$\boxed{x_A/x_B = 0.429}$$

Problem 7.5

Due to its geographical location, Portland, Oregon, USA can experience a wide range of weather patterns during the winter. If wind blows in from the ocean to the west, the temperature is warmer but generally brings ocean moisture which causes rain. If wind comes from Canada to the north or from the mountains/plans to the east, the weather is cold but dry. If the wind comes from California to the south, the temperature is warm and dry. Given the surface weather maps below, utilize your knowledge of fluid behavior and pressure to predict the resulting weather in Portland.

a) High Pressure system to the East

b) Low pressure system to the North

c) High Pressure system to the South

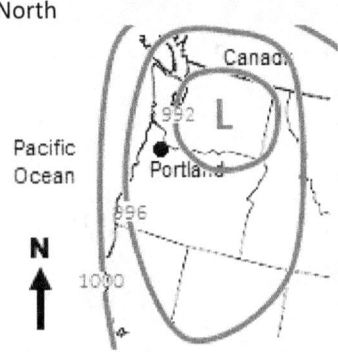
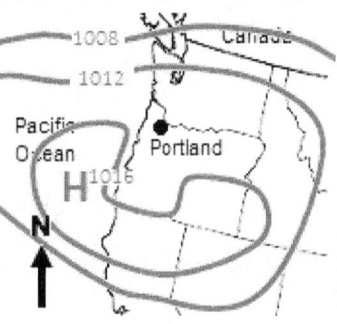

a) Air will flow away from the high pressure system. A high pressure system to the east of Portland will cause winds from the east. As such, the weather is expected to be cold and dry.

b) Air will flow towards the low pressure system. A low pressure system to the north of Portland will cause air to travel north, so in Portland the wind will come from the south. As such, the weather is expected to be warm and dry.

c) This high pressure system will cause air to flow from both the south and from the west towards Portland. The resulting weather will be a combination of the southern warmth and the ocean moisture. As such, the weather in Portland will be warm and wet.

Problem 7.6

Calculate the pressure produced by a force of 800 N acting on an area of 2.0 m².

$P = F/A$
$P = 800 \text{ N} / 2.0 \text{ m}^2$
$P = 400 \text{ N} / \text{m}^2$

$\boxed{P = 400 \text{ Pa}}$

Problem 7.7
Calculate the pressure due to a column of water at depths of 10 meters, 50 meters, and 100 meters.

We are concerned only with the pressure due to the liquid column, so $P = \rho g h$.

At 10 meters:
$P = (1000 \text{ kg/m}^3)(9.8 \text{ m/s}^2)(10 \text{ m})$
$P = 98,000 \text{ N/m}^2$

$\boxed{P = 9.8 \times 10^5 \text{ N/m}^2}$

At 50 meters:
$P = (1000 \text{ kg/m}^3)(9.8 \text{ m/s}^2)(50 \text{ m})$
$P = 98,000 \text{ N/m}^2$

$\boxed{P = 490,000 \text{ N/m}^2}$

At 100 meters:
$P = (1000 \text{ kg/m}^3)(9.8 \text{ m/s}^2)(100 \text{ m})$
$P = 98,000 \text{ N/m}^2$

$\boxed{P = 9.8 \times 10^6 \text{ N/m}^2}$

Problem 7.8
Given the oil + water system below, what is the pressure at the liquid interface? What is the pressure at the bottom of the column?

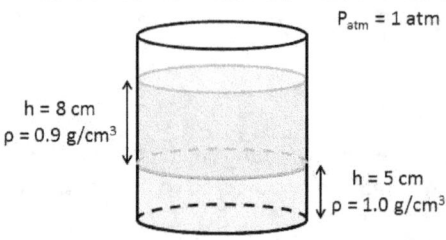

The pressure at the interface is due only to the height of the oil, so $P = P_0 + \rho_{oil} \, g \, h_{oil}$:
$P = 101325 \text{ N/m}^2 + (900 \text{ kg/m}^3)(9.8 \text{ m/s}^2)(0.08 \text{ m})$
$P = 101325 \text{ N/m}^2 + 705.6 \text{ N/m}^2$
$P = 102030.6 \text{ N/m}^2$

$\boxed{P_{interface} = 102.0 \text{ kPa}}$

The pressure at the interface is due only to the height of the oil, so $P = P_0 + \rho_{oil} \, g \, h_{oil} + \rho_{water} \, g \, h_{water}$:
$P = 101325 \text{ N/m}^2 + (900 \text{ kg/m}^3)(9.8 \text{ m/s}^2)(0.08 \text{ m}) + (1000 \text{ kg/m}^3)(9.8 \text{ m/s}^2)(0.05 \text{ m})$
$P = 101325 \text{ N/m}^2 + 705.6 \text{ N/m}^2 + 490 \text{ N/m}^2$
$P = 102520.6 \text{ N/m}^2$

$\boxed{P_{interface} = 102.5 \text{ kPa}}$

Problem 7.9

Two equal-sized tanks are connected. One contains water, and the other contains oil with density of 0.9 g/cm³. Both tanks are filled to an initial liquid height of 3 meters. When the valve is opened, what will be the final liquid level height in each tank? Note the oil and water do not mix.

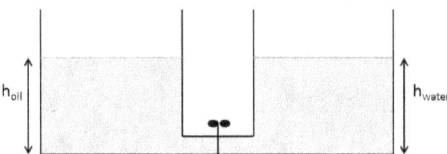

The water will move into the oil tank to compensate for the low pressure in that tank. At equilibrium, pressure at the bottom of both tanks is the same:

$$P_1 = P_2$$

$$\rho_{oil}\, g\, h_{oil} + \rho_{water}\, g\, h_{water,tank1} = \rho_{water}\, g\, h_{water,tank2}$$

$$\rho_{oil}\, h_{oil} + \rho_{water}\, h_{water,tank1} = \rho_{water}\, h_{water,tank2}$$

Because the total amount of water remains the same, $\Delta h_1 = -\Delta h_2$.

$$\rho_{oil}\, h_{oil} + \rho\, h_{water,tank1} = \rho_{water}\,(h_{water,initial} - h_{water,tank1})$$

Then solving for the height of water in tank 1:

$$\rho_{oil}\, h_{oil} + \rho\, h_{water,tank1} = \rho_{water}\,(h_{water,initial} - h_{water,tank1})$$

$$\rho_{oil}\, h_{oil} + \rho\, h_{water,tank1} = \rho_{water}\, h_{water,initial} - \rho_{water}\, h_{water,tank1}$$

$$\rho_{oil}\, h_{oil} + 2\,\rho\, h_{water,tank1} = \rho_{water}\, h_{water,tank2}$$

$$\frac{2\,\rho_{water}\, h_{water,initial} - \rho_{oil}\, h_{oil}}{2\,\rho_{water}} = h_{water,tank1}$$

$$\frac{900\ \text{kg/m}^3\ \cdot 3\text{m} + 1000\ \text{kg/m}^3\ \cdot 3\text{m}}{2 \cdot 1000\ \text{kg/m}^3} = h_{water,tank1}$$

$$0.15\ \text{m} = h_{water,tank1}$$

Now solving for the final tank heights:
Tank 1 = 3m + 0.15 m = 3.15 m
Tank 2 = 3m − 0.15 m = 2.85 m

> Tank 1 has 3.15 m of liquid, and tank 2 has 2.85 m of liquid

Problem 7.10

Given the system below, what will be the final height of the liquid in tank 1 once the valve is opened? The height in tank 2?

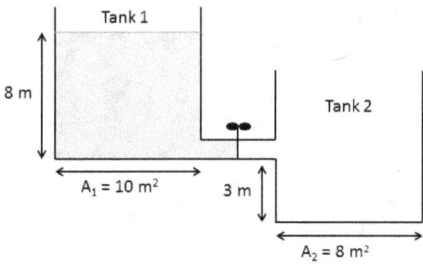

Let us first declare our variables:
h_1 – initial height in tank 1
$h_{1,final}$ – final height in tank 1
h_2 – final height in tank 1 above the crossover
h_0 – height in tank 2 below the crossover

At equilibrium, the pressure in both tanks above the cross-over is the same:

$$P_1 = P_2$$
$$\cancel{\rho g}\, h_{1,final} = \cancel{\rho g}\, h_2$$
$$h_{1,final} = h_2$$

The total volume of liquid remains the same, so:

volume initial = volume final

initial volume in tank 1 = final volume in tank 1 + volume in tank 2

$$h_1 A_1 = h_{1,final} A_1 + A_2 (h_2 + h_0)$$

Substitute for $h_{1,final} = h_2$, and then solving for $h_{1,final}$:

$$h_1 A_1 = h_{1,final} A_1 + A_2 (h_{1,final} + h_0)$$
$$h_1 A_1 = h_{1,final} A_1 + A_2 h_{1,final} + A_2 h_0$$
$$h_1 A_1 = (A_1 + A_2) h_{1,final} + A_2 h_0$$
$$h_{1,final} = \frac{A_1 h_1}{(A_1 + A_2)} - \frac{A_2 h_0}{(A_1 + A_2)}$$
$$h_{1,final} = \frac{10 \text{ m}^2 \cdot 8 \text{ m}}{(10 \text{ m}^2 + 8 \text{ m}^2)} - \frac{8 \text{ m}^2 \cdot 3 \text{ m}}{(10 \text{ m}^2 + 8 \text{ m}^2)}$$
$$h_{1,final} = \frac{80 \text{ m}^3}{(18 \text{ m}^2)} - \frac{24 \text{ m}^3}{(18 \text{ m}^2)} = 3.11 \text{ m}$$

Solving for the total height in tank 2:
$h_{tank2} = h_2 + h_0$
$h_{tank2} = 3.11\ m + 3\ m = 6.11\ m$

> The final height in tank 1 is 3.11 m
> The final height in tank 2 is 6.11 m

Problem 7.11
Given the system below, what will be the final height of the liquid in tank 1 once the valve is opened? The height in tank 3?

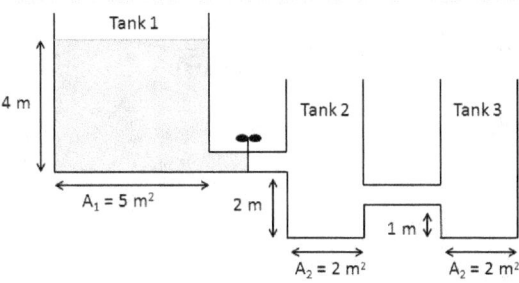

Let us first declare our variables:
h_1 – initial height in tank 1
$h_{1,final}$ – final height in tank 1
h_2 – final height in tank 1 above the tank1/2 crossover
$h_{2,0}$ – height in tank 2 below the tank2/3 crossover
$h_{2,gap}$ – gap in tank 2 between the two crossovers (= 1m)
$h_{3,0}$ – height in tank 3 below the tank2/3 crossover

At equilibrium, the pressure in both Tank 1 and Tank 2 above the cross-over is the same:
$$P_1 = P_2$$
$$\cancel{\rho g}\, h_{1,final} = \cancel{\rho g}\, h_2$$
$$h_{1,final} = h_2$$

At equilibrium, the pressure in both tanks above the cross-over is the same:
$$P_2 = P_3$$
$$\cancel{\rho g}\,(h_{2,gap} + h_2) = \cancel{\rho g}\, h_3$$
$$h_{2,gap} + h_2 = h_3$$

The total volume of liquid remains the same, so:
$$\text{volume initial} = \text{volume final}$$
$$\text{initial volume in tank 1} = \text{final volume in tank 1} + \text{volume in tank 2} + \text{volume in tank 3}$$
$$h_1 A_1 = h_{1,final} A_1 + A_2 (h_{2,0} + h_{2,gap} + h_2) + A_3 (h_{3,0} + h_3)$$

Substitute for h_2 and h_3 and then solving for $h_{1,final}$:

$$h_1 A_1 = h_{1,final} A_1 + A_2 (h_{1,final} + h_{2,0} + h_{2,gap}) + A_3 (h_{3,0} + h_{2,gap} + h_{1,final})$$

$$h_1 A_1 = h_{1,final} A_1 + A_2 h_{1,final} + A_2 h_{2,0} + A_2 h_{2,gap} + A_3 h_{3,0} + A_3 h_{2,gap} + A_3 h_{1,final}$$

$$h_1 A_1 = (A_1 + A_2 + A_3) h_{1,final} + (A_1 + A_2) h_{2,gap} + A_2 h_{2,0} + A_3 h_{3,0}$$

$$h_{1,final} = \frac{A_1 h_{1,final}}{(A_1 + A_2 + A_3)} - \frac{(A_2 + A_3) h_{2,gap}}{(A_1 + A_2 + A_3)} - \frac{A_2 h_{2,0}}{(A_1 + A_2 + A_3)} - \frac{A_3 h_{3,0}}{(A_1 + A_2 + A_3)}$$

$$h_{1,final} = \frac{5 \text{ m}^2 \cdot 4 \text{ m}}{(5 \text{ m}^2 + 2 \text{ m}^2 + 2 \text{ m}^2)} - \frac{(2 \text{ m}^2 + 2 \text{ m}^2) \cdot 1 \text{ m}}{(5 \text{ m}^2 + 2 \text{ m}^2 + 2 \text{ m}^2)} - \frac{2 \text{ m}^2 \cdot 1 \text{ m}}{(5 \text{ m}^2 + 8 \text{ m}^2 + 2 \text{ m}^2)}$$
$$- \frac{2 \text{ m}^2 \cdot 1 \text{ m}}{(5 \text{ m}^2 + 8 \text{ m}^2 + 2 \text{ m}^2)}$$

$$h_{1,final} = \frac{20 \text{ m}^2}{(9 \text{ m}^2)} - \frac{4 \text{ m}^2}{(9 \text{ m}^2)} - \frac{2 \text{ m}^3}{(9 \text{ m}^2)} - \frac{2 \text{ m}^3}{(9 \text{ m}^2)} = \mathbf{1.33 \text{ m}}$$

Solving for the total height in tank 3:
$h_{tank3} = h_3 + h_{3,0}$
$h_{tank3} = h_{2,gap} + h_2 + h_{3,0}$
$h_{tank3} = h_{2,gap} + h_{1,final} + h_{3,0}$
$\mathbf{h_{tank2} = 1 \text{ m} + 1.33 + 1 \text{ m} = 3.33 \text{ m}}$

> The final height in tank 1 is 1.33 m
> The final height in tank 3 is 3.33 m

Problem 7.12
A hydraulic lift has a 10 kg weight supported on a circular piston head with radius of 10 cm, and a 1,000 kg weight supported on a circular piston head with radius of 1 m. Which side of the lift moves upward?

The pressure on the 10 kg weight side is:
P = F/A
P = mg / A
P = 10 kg · 9.8 m/s² / π (0.01 m)²
P = 311,943 N/m²

The pressure on the 1000 kg weight side is:
P = F/A
P = mg / A
P = 1000 kg · 9.8 m/s² / π (1 m)²
P = 3,119.43 N/m²

> 1000 kg weight will move upwards

Chapter 7 | Fluid Mechanics

Problem 7.13

A hydraulic lift has a circular piston head with radius of 20 cm, and a 2,000 kg car supported on a rectangular piston head with width of 2 m and length of 3 m. How far must the small piston be pressed down in order to lift the car by 2 m?

Start by finding the area of the two sides of the lift:
$A_{small} = \pi r^2 = \pi (0.02 \text{ m})^2 \rightarrow A_{small} = 1.256 \times 10^{-3} \text{ m}^2$
$A_{large} = l \cdot w = 2 \text{ m} \cdot 3 \text{ m} \rightarrow A_{large} = 6 \text{ m}^2$

We can use the equation relating distance in a hydraulic lift:
$$d_{small} = d_{large} \left(\frac{A_{large}}{A_{small}} \right)$$
$$d_1 = 2m \left(\frac{6 \text{ m}^2}{1.256 \times 10^{-3} \text{ m}^2} \right)$$
$$d_1 = 9{,}554.1 \text{ m}$$

> We need to press the small piston down by 9,444.1 meters.
> Obviously, we should select a larger piston head!

Problem 7.14

A column of mercury is open to the atmosphere on a day when the atmospheric pressure is 29.6 in Hg. What is the gauge pressure 2 in below the surface? The absolute pressure? Give answers in in Hg.

The pressure 2 inches below the surface is 2 in Hg. We can then solve for the absolute pressure:
$P_{abs} = P_{surface} + P_{gauge}$
$P_{abs} = 29.6 \text{ in Hg} + 2 \text{ in Hg}$

> $P_{gauge} = 2 \text{ in Hg}$
> $P_{abs} = 31.6 \text{ in Hg}$

Problem 7.15

Visitors to Crater Lake National Park in Oregon, USA, can drive around the rim of a crater formed thousands of years ago when a volcano collapsed inwards after an eruption. Within the crater lies the deepest lake in the United States, Crater Lake, which has its deepest point at 1943 ft (~592 m) below the surface. Because it is located high in the mountains, Crater Lake has an average surface elevation of 6173 ft (~1881.5 m) above sea level.
1) What is the gauge pressure (in atm) at Crater Lake's deepest point?
2) Estimate the pressure (in atm) due to the atmosphere at the lake surface assuming the density of air remains constant.
3) Using your above answers, find the absolute pressure (in atm) at Crater Lake's deepest point.

The gauge pressure at the bottom of the lake is due simply to the column of water:

$P = \rho g h$

$P = (1000 \text{ kg/m}^3)(9.8 \text{ m/s}^2)(592 \text{m}) \cdot (1 \text{ atm}/101325 \text{ Pa})$

$P = 5,801,600 \text{ Pa} \cdot (1 \text{ atm}/101325 \text{ Pa})$

$P = 57.25 \text{ atm}$

$$\boxed{P_{gauge} = 57.25 \text{ atm}}$$

Because we are increasing in elevation, to find the pressure due to atmosphere at the higher elevation, we subtract the pressure due to the column of air from the pressure at sea level:

$P_{surface} = P_{atm} - \rho g h$

$P_{surface} = 1 \text{ atm} - (1.2 \text{ kg/m}^3)(9.8 \text{ m/s}^2)(1881.5 \text{ m}) \cdot (1 \text{ atm}/101325 \text{ Pa})$

$P_{surface} = 1 \text{ atm} - (1.2 \text{ kg/m}^3)(9.8 \text{ m/s}^2)(1881.5 \text{ m}) \cdot (1 \text{ atm}/101325 \text{ Pa})$

$P_{surface} = 1 \text{ atm} - 0.218 \text{ atm}$

$$\boxed{P_{surface} = 0.782 \text{ atm}}$$

We can now solve for the absolute pressure at the bottom of the lake:

$P_{abs} = P_{surface} + P_{gauge}$

$P_{abs} = 57.25 \text{ atm} + 0.782 \text{ atm}$

$$\boxed{P_{abs} = 58.03 \text{ atm}}$$

Problem 7.16

Express a head pressure of 200 kPa in terms of mmHg.

Mercury has a density of 13,600 kg/m³.

$$P = \rho g h \rightarrow h = \frac{P}{\rho g}$$

| $\frac{200 \times 10^3 \text{ N}}{\text{m}^2}$ | $\frac{\text{m}^3}{13{,}600 \text{ kg}}$ | $\frac{\text{s}^2}{9.8 \text{ m}}$ | $\frac{1 \text{ kg m/s}^2}{\text{N}}$ | $\frac{10^3 \text{ mm}}{\text{m}}$ | $= 1.5 \times 10^3 \text{ mmHg}$ |

$$\boxed{P = 1.5 \times 10^3 \text{ mmHg}}$$

Problem 7.17

Express a head pressure of 200 kPa in terms of inHg.

Mercury has a density of 1,000 kg/m³.

$$P = \rho g h \rightarrow h = \frac{P}{\rho g}$$

| $\frac{200 \times 10^3 \text{ N}}{\text{m}^2}$ | $\frac{\text{m}^3}{1{,}000 \text{ kg}}$ | $\frac{\text{s}^2}{9.8 \text{ m}}$ | $\frac{1 \text{ kg m/s}^2}{\text{N}}$ | $\frac{10^3 \text{ mm}}{\text{m}}$ | $= 20.4 \times 10^3 \text{ mmH}_2\text{O}$ |

$$\boxed{P = 20.4 \times 10^3 \text{ mmH}_2\text{O}}$$

Problem 7.18

The pressure in a vessel filled with an ideal gas is measured by an open-end manometer as shown below. The temperature of the gas is originally 20 °C. If the temperature of the vessel is raised to 50 °C, what is the new manometer fluid height?

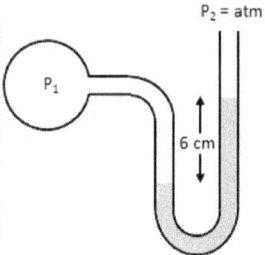

We start by putting everything into SI units:
6 cm = 0.06 m
1 atm = 101.3 kPa = 101325 N/m²

Next we find the initial pressure in the vial:
$P_1 = P_{atm} + \rho g h$
P_1 = 101325 N/m² + (13550 kg/m³)(9.8 m/s²)(0.06 m)
P_1 = 109292.4 N/m²

We can then use the ideal gas law to find the new pressure at 50 °C (there is no significant change in container volume, so it reduces to P/T):
$P_1 / T_1 = P_2 / T_2$
$P_2 = P_1 T_2 / T_1$
P_2 = (109292.4 N/m²) (323 K/293 K)
P_2 = 120482.7 N/m²

Finally, we solve for the new height of mercury:
$P_2 = P_{atm} + \rho g h$
$(P_2 - P_{atm}) / \rho g = h$
h = (120482.7 N/m² − 101325 N/m²) / (13550 kg/m³)(9.8 m/s²)
h = 0.144 m

h = 14.4 cm

Problem 7.19

Two 1 L vessels filled with ideal gases are connected by a differential manometer filled with mercury as shown below. Both vessels are originally at 20 °C. If the temperature of both vessels is then raised to 50 °C, what is the new pressure difference them?

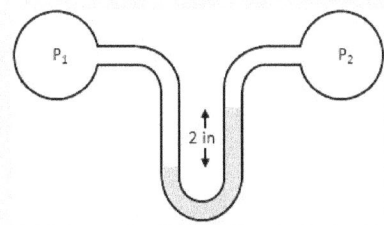

2 in = 0.0508 m
1 L = 0.001 m³
R = 8.314 Nm/mol K

If the two containers of the same volume are at different pressures when at the same temperature, then we can conclude that there are more moles of gas in container 1. Thus, at the new temperature, we expect to see a greater increase in P_1 than in P_2. The first step is to solve for n_1 and n_2, the moles in each container.

$$P_1 = \frac{n_1 RT}{V} \quad \text{and} \quad P_2 = \frac{n_2 RT}{V}$$

However, we do not know the exact pressures in each container, which makes it difficult to solve for the exact number of moles in each container. We do, however, know the difference between them - the difference in pressures is given by the manometer fluid height:

$$P_1 - P_2 = \rho g h$$
$$\frac{n_1 RT}{V} - \frac{n_2 RT}{V} = \rho g h$$
$$(n_1 - n_2)\frac{RT}{V} = \rho g h$$
$$\Delta n = \frac{\rho g h V}{RT}$$

Solving for the difference (Δn):

$$\Delta n = \frac{\rho g h V}{RT} = \frac{(13550 \frac{kg}{m^3})(9.8 \frac{m}{s^2})(0.0508\ m)(0.001\ m^3)}{(8.314 \frac{N\ m}{mol\ K})(293\ K)}$$

Δn = 2.769 x 10³ mol

The difference in moles will lead to the change in pressure between the two vessels as the temperature goes up:

$$\Delta P = \frac{\Delta n\ R\ \Delta T}{V}$$

$$\Delta P = \frac{(2.769 \times 10^3\ mol)(8.314 \frac{N\ m}{mol\ K})(30\ K)}{0.001\ m^3}$$

ΔP = 690.64 N/m²

The total pressure between them is then:

$\Delta P_F = \Delta P_0 + \Delta P$

$\Delta P_F = \rho g h + \Delta P$

$\Delta P_F = (13550 \text{ kg/m}^3)(9.8 \text{ m/s}^2)(0.0508\text{m}) + 690.64 \text{ N/m}^2$

$\Delta P_F = 7436.37 \text{ N/m}^2$

$$\boxed{\Delta P_F = 7.43 \text{ kPa}}$$

Problem 7.20

Two vessels are connected by a differential manometer filled with mercury as shown below. If the pressure in both vessels is the same, determine the density of the fluid in the vessels.

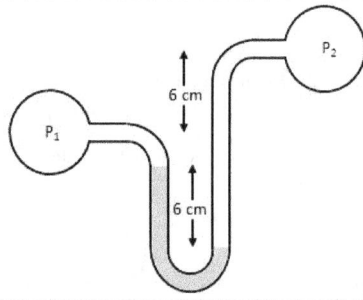

This is a concept problem because no calculations are needed.

The pressure in the vessels is the same, thus all of the pressure difference in the manometer fluid comes from the height difference between the two vessels. Since the height difference and the height of the manometer fluid is the same, the density of the fluid in the vessels is the same as the manometer fluid.

The fluid in the vessels is mercury!

Problem 7.21

Fluid flowing through a pipe is being measured by an open-end manometer filled with mercury. A reading of 4 cm is obtained. What is the gauge pressure in mmHg? What is the absolute pressure in mmHg?

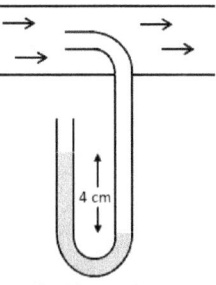

$P_{gauge} = \rho\, g\, h$
$P_{gauge} = (13550 \text{ kg/m}^3)(9.8 \text{ m/s}^2)(0.04 \text{ m})$
$P_{gauge} = 5311.6 \text{ N/m}^2$

$$\boxed{\Delta P_{gauge} = 5.3 \text{ kPa}}$$

$P_{abs} = P_{atm} + P_{gauge}$
$P_{abs} = 101325 \text{ N/m}^2 + 5311.6 \text{ N/m}^2$
$P_{abs} = 106636.6 \text{ N/m}^2$

$$\boxed{P_{abs} = 106.6 \text{ kPa}}$$

Problem 7.22

A floating object displaces 0.5 m³ of water. Calculate the buoyant force on the object and the weight of the object.

We can calculate the buoyant force from the displaced water:
$F_{buoyancy} = \rho_{liquid}\, g\, V_{displaced}$
$F_{buoyancy} = (1000 \text{ kg/m}^3)(9.8 \text{ m/s}^2)(0.5 \text{ m}^3)$
$F_{buoyancy} = 4900 \text{ N}$

$$\boxed{F_{buoyancy} = 4900 \text{ N}}$$

Because the object is floating, the net force on the object is zero. Thus, the weight of the object is the same as the buoyant force.

$$\boxed{W = 4900 \text{ N}}$$

Problem 7.23

A block of wood of mass 4 kg floats in water. What is the buoyant force on the block?

The wooden block is floating, so the buoyant force must equal the weight of the object.

$F_{buoyancy} = W = m\,g$

$F_{buoyancy} = (4 \text{ kg})(9.8 \text{ m/s}^2)$

$F_{buoyancy} = 39.2 \text{ N}$

$$\boxed{F_B = 39.2 \text{ N}}$$

Problem 7.24

A 20 kg boy has built a wooden raft that is 1 meter wide, 3 meters long, and 0.5 m high. If the wood is plywood with density of 0.546 g/cm³, will the boy remain dry?

The depth to which a rectangular floating object will sink was derived to be:

$$d = \frac{m_{object}}{\rho_{liquid}\; lw}$$

The mass of the raft and the boy is:

$m_{object} = m_{boy} + m_{raft}$

$m_{object} = 20 \text{ kg} + (546 \text{ kg/m}^3)(1\text{m} \cdot 3\text{m} \cdot 0.5\text{m})$

$m_{object} = 20 \text{ kg} + 819 \text{ kg}$

$m_{object} = 839$ kg

Now plugging into the equation:

$$d = \frac{m_{object}}{\rho_{liquid}\; lw}$$

$$d = \frac{839 \text{ kg}}{(1000\,\frac{\text{kg}}{\text{m}^3})(1\text{m} \cdot 3\text{m})}$$

$d = 0.280$ m

$$\boxed{\text{The raft will sink by 0.28 meters, and it is 0.5 high. Thus, the boy will remain dry!}}$$

Problem 7.25

How much mass can one cubic meter of helium lift?

The buoyant force on the helium is:
B = V_{disp} ρ_{air} g
B = $(1 \text{ m}^3)(1.200 \text{ kg/m}^3)(9.8 \text{ m/s}^2)$
B = 11.76 N

The weight of the helium itself is:
W = m_{He} g
W = V_{He} ρ_{He} g
W = $(1 \text{ m}^3)(0.179 \text{ kg/m}^3)(9.8 \text{ m/s}^2)$
W = 1.7542 N

So the force left to lift is:
F_{up} = B − W = 11.76 N − 1.7542 N
F_{up} = 10.0058 N

Solving for the mass that force can lift:
m = F_{up} / g = (10.0058 N) / (9.8 m/s^2)
m = 1.021 kg

> **1 m³ of Helium can lift approximately 1.02 kg**
>
> *Note: this means that, approximately, 1 L of Helium can lift 1 gram of material.*

Problem 7.26

"It's just the tip of the iceberg". What fraction of an iceberg made of pure ice is submerged below the surface of the salt water ocean?

The weight of the iceberg is:
$W = m_{ice}\, g$

The buoyant force is:
$B = \rho_{sea}\, V_{sea}\, g$

The volume of the iceberg is:
$V_{ice} = m_{ice} / \rho_{ice}$ → $m_{ice} = V_{ice}\, \rho_{ice}$

Ice has a specific gravity of 0.917, and sea water has a specific gravity of 1.025. Thus, the iceberg is floating on the ocean surface. Because it is floating, we know that B = W. Equating the two:

$B = W$
$\rho_{sea}\, V_{sea}\, g = m_{ice}\, g$
$\rho_{sea}\, V_{sea} = m_{ice}$

Substituting in for the volume of ice:
$\rho_{sea}\, V_{sea} = V_{ice}\, \rho_{ice}$

$$\frac{V_{sea}}{V_{ice}} = \frac{\rho_{ice}}{\rho_{sea}}$$

The volume of ice that is submerged is equal to the volume of displaced water. Solving the ratio of the volume of water to the entire volume of the iceberg gives the % that is submerged:

$$\frac{V_{sea}}{V_{ice}} = \frac{\rho_{ice}}{\rho_{sea}} = \frac{0.917}{1.025}$$

$$\frac{V_{sea}}{V_{ice}} = 0.894$$

89.4% of the iceberg is below the surface

Problem 7.27

You collect 12 kg of water in two minutes coming out of a pipe. What is the volumetric flow rate of the water through a pipe with diameter of 1cm? What is the volumetric flow rate through a pipe with diameter of 5 cm?

The volumetric flow rate is:

$$\dot{m} = \frac{mass}{time} = \frac{12\ kg}{3\ min} = 4\ kg/min$$

$$\boxed{\dot{m} = 4\ L/min}$$

Problem 7.28

You collect 6 L of water in two minutes coming out of a pipe. What is the volumetric flow rate of the water through a pipe with diameter of 1cm? What is the volumetric flow rate through a pipe with diameter of 5 cm?

This is somewhat of a trick question – the volumetric flow rate will not depend on the size of the pipe. The fluid velocity will change, but not the volumetric flow rate.

The volumetric flow rate is:

$$\dot{V} = \frac{Volume}{time} = \frac{6\ L}{2\ min} = 3\ L/min$$

$$\boxed{\dot{V} = 3\ L/min\ \text{(regardless of pipe diameter)}}$$

Problem 7.29

The mass flow rate of a liquid with density 0.92 g/cm³ in a pipe is 50 kg/min. What is the volumetric flow rate of the fluid?

We solve for the volumetric flow rate using:

$$\dot{V} = \frac{\dot{m}}{\rho} = \frac{50.0\ kg/min}{920\ kg/m^3}$$
$$\dot{V} = 0.054\ m^3/min$$

$$\boxed{\dot{V} = 0.054\ m^3/min}$$

Problem 7.30

The volumetric flow rate of a liquid with density 1.2 g/cm3 in a pipe is 400 cm3/min. What is the mass flow rate of the fluid?

We solve for the mass flow rate using:

$$\dot{m} = \frac{\dot{V}}{\rho} = \frac{400\ cm^3/min}{1.2\ g/cm^3}$$
$$\dot{m} = 333.3\ g/min$$

$$\boxed{\dot{m} = 0.333\ kg/min}$$

Problem 7.31

If a fluid with density of 1.5 g/cm³ is flowing through a 3 cm diameter pipe at 10.0 kg/min, what is the average velocity of the bulk fluid?

First we need to solve for the volumetric flow rate:

$$\rho = \frac{\dot{m}}{\dot{V}}$$

$$\dot{V} = \frac{\dot{m}}{\rho} = \frac{10.0 \text{ kg/min}}{1500 \text{ kg/m}^3}$$

$$\dot{V} = 6.66 \times 10^{-3} \text{ m}^3/\text{min}$$

Then we use the volumetric flow rate and the pipe size to solve for the bulk fluid velocity:

$$v = \frac{\dot{V}}{A}$$

$$v = \frac{\dot{V}}{\pi \left(\frac{d}{2}\right)^2}$$

$$v = \frac{6.66 \times 10^{-3} \text{ m}^3/\text{min}}{\pi \left(\frac{0.03 \text{ m}}{2}\right)^2}$$

$$v = \frac{6.66 \times 10^{-3} \text{ m}^3/\text{min}}{7.06 \times 10^{-4} \text{ m}^2}$$

v = 9.43 m/min

$$\boxed{v = 9.43 \text{ m/min}}$$

Problem 7.32

For intravenous (IV) drug delivery, medical professionals set the drug flow rate using a measurement called the 'drop rate', which is given in drops/minute. The standard conversion is that there are 15 drops per mL. A doctor will order an amount of drug be delivered to a patient over a specific duration, and it is necessary to then determine the appropriate drop rate to meet that order. If a doctor orders 1500 mL of saline over a 12 hour period, what is the flow rate in SI units? What is the drop rate?

The flow rate is the volume per time:

$$\dot{V} = \frac{volume}{time} = \frac{1500\ mL}{12\ hours}$$

$$\dot{V} = \frac{1500\ mL}{12\ hours} \cdot \frac{m^3}{1 \times 10^6\ mL} \cdot \frac{hr}{3600\ s}$$

$$\dot{V} = 3.47 \times 10^{-8}\ m^3/s$$

$$\boxed{\dot{V} = 3.47 \times 10^{-8}\ m^3/s}$$

Drop rate:

$$\dot{V} = \frac{1500\ mL}{12\ hours} \cdot \frac{15\ drop}{mL} \cdot \frac{hr}{60\ min}$$

$$\dot{V} = 31.25\ drops/min$$

$$\boxed{\dot{V} = 31.25\ \text{drops/min}}$$

Problem 7.33

In the figure below, in which region (A, B, or C) would you expect the pressure on the walls of the pipe to be greatest?

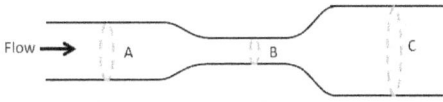

For pressure:
Region C > Region A > Region B

For fluid velocity:
Region B > Region A > Region C

$$\boxed{\text{Region C has the greatest pressure}}$$

Problem 7.34

Water flows through a 1 inch diameter hose with a bulk velocity of 2 ft/second. Find the speed of the water in ft/s through a nozzle with diameter 1/8th of an inch.

The cross-sectional area of the hose is:
$A_1 = \pi r^2$
$A_1 = \pi (0.5 \text{ in})^2$
$A_1 = 0.785 \text{ in}^2$

And the cross-sectional area of the exit nozzle is:
$A_2 = \pi r^2$
$A_2 = \pi (0.125 \text{ in})^2$
$A_2 = 0.049 \text{ in}^2$

This is a simple continuity problem:

$$A_1 v_1 = A_2 v_2$$
$$(0.785 \text{ in}^2)(2 \text{ ft/s}) = (0.125 \text{ in}^2)(v_2)$$
$$12.56 \text{ ft/s} = v_2$$

$$\boxed{v_2 = 12.56 \text{ ft/s}}$$

Problem 7.35

Water is flowing through a hose with velocity of 1.0 m/s and a pressure of 200 kPa. The end of the hose is at atmospheric pressure (101.325 kPa), and there is no significant change in height. What is the velocity of the water exiting the hose?

This is a classic Bernoulli equation problem. There is no change in height, so the potential energy terms cancel:

$$P_1 + \frac{1}{2}\rho v_1^2 + \rho g h_1 = P_2 + \frac{1}{2}\rho v_2^2 + \rho g h_2$$

$$P_1 + \frac{1}{2}\rho v_1^2 + \cancel{\rho g h_1} = P_2 + \frac{1}{2}\rho v_2^2 + \cancel{\rho g h_2}$$

Solving then for the exit velocity, v_2:

$$P_1 - P_2 + \frac{1}{2}\rho v_1^2 = \frac{1}{2}\rho v_2^2$$

$$\frac{2}{\rho}\left(P_1 - P_2 + \frac{1}{2}\rho v_1^2\right) = v_2^2$$

$$\sqrt{\frac{2}{\rho}\left(P_1 - P_2 + \frac{1}{2}\rho v_1^2\right)} = v_2$$

$$\sqrt{\frac{2}{1000 \text{ kg/m}^2}\left(200{,}000 \text{ Pa} - 101{,}325 \text{ Pa} + \frac{1}{2}(1000 \text{ kg/m}^2)(1.0 \text{ m/s})^2\right)} = v_2$$

$$\boxed{v_2 = 14.08 \text{ m/s}}$$

Problem 7.36

Propane (density is 495 kg/m³) is burned as fuel to create steam for a power plant. Propane is flowing through a pipe at a velocity of 1 m/s and 101.325 kPa, but must be delivered to the combustion unit at a pressure of 200 kPa on a lower level. How far must the pipe drop in height in order to achieve the desired pressure? Assume velocity is constant.

This is a classic Bernoulli equation problem. There is no change in velocity, so the kinetic energy terms cancel:

$$P_1 + \frac{1}{2}\rho v_1^2 + \rho g h_1 = P_2 + \frac{1}{2}\rho v_2^2 + \rho g h_2$$

$$P_1 + \cancel{\frac{1}{2}\rho v_1^2} + \rho g h_1 = P_2 + \cancel{\frac{1}{2}\rho v_2^2} + \rho g h_2$$

Solving then for the change in height ($h_2 - h_1$):

$$\rho g h_1 - \rho g h_2 = P_2 - P_1$$

$$\rho g (h_1 - h_2) = P_2 - P_1$$

$$\Delta h = \frac{P_2 - P_1}{\rho g}$$

$$\Delta h = \frac{200{,}000 \text{ Pa} - 101325 \text{ Pa}}{(495 \text{ kg/m}^3)(9.81 \text{ m/s}^2)}$$

$$\boxed{\Delta h = 20.3 \text{ m}}$$

Problem 7.37

A fountain designed to spray a column of water 10 m into the air has a 1 cm diameter nozzle at ground level. The water pump is 3 m below the ground, and feeds through a 2 cm diameter pipe to the nozzle. Find the pump pressure necessary if the fountain is to operate as designed.

We start with the Bernoulli equation, and solve for P_1 (the pump pressure). Because the fountain nozzle is being chosen as the reference height, this makes the change from h_2 to h_1 a negative value:

$$P_1 + \frac{1}{2}\rho v_1^2 + \rho g h_1 = P_2 + \frac{1}{2}\rho v_2^2 + \rho g h_2$$

$$P_1 = P_2 + \frac{1}{2}\rho v_2^2 + \rho g h_2 - \frac{1}{2}\rho v_1^2 - \rho g h_1$$

$$P_1 = P_2 + \frac{1}{2}\rho v_2^2 - \frac{1}{2}\rho v_1^2 - \rho g (h_1 - h_2)$$

Next, we start by identifying the variables for the Bernoulli equation. The pressure at ground level is 1 atm, or 101325 N/m². So $P_2 = 101325$ N/m².

To find v_2, we must determine the velocity at which the water must leave the nozzle in order to reach the desired height. At the top of the column, the velocity is zero (before the water crashes back down):

$$\begin{aligned}
v_f^2 &= v_i^2 + g h \\
0 &= v_{nozzle}^2 + g h \\
v_{nozzle} &= \sqrt{g h} \\
v_{nozzle} &= \sqrt{(9.81 \text{ m/s}^2)(10 \text{ m})} \\
v_{nozzle} &= 9.90 \text{ m/s}
\end{aligned}$$

V_1 is related to v_2 by the continuity equation:

$$\begin{aligned}
A_1 v_1 &= A_2 v_2 \\
v_1 &= (A_2/ A_1) v_2 \\
v_1 &= (\pi r_2^2 / \pi r_1^2) v_2 \\
v_1 &= ((0.5 \text{ cm})^2 / (1 \text{ cm})^2) v_2 \\
v_1 &= 0.25 v_2
\end{aligned}$$

So now plugging into the Bernoulli equation:

$$P_1 = 101325 \text{ N/m}^2 + \frac{1}{2}\left(1000 \frac{\text{kg}}{\text{m}^3}\right)\left(9.90 \frac{\text{m}}{\text{s}}\right)^2 - \frac{1}{2}\left(1000 \frac{\text{kg}}{\text{m}^3}\right)\left(0.25 \cdot 9.90 \frac{\text{m}}{\text{s}}\right)^2 - \left(1000 \frac{\text{kg}}{\text{m}^3}\right)\left(9.81 \frac{\text{m}}{\text{s}^2}\right)(-3\text{m})$$

$$P_1 = 101{,}325 \text{ N/m}^2 + 49{,}005 \text{ N/m}^2 - 3{,}062.8 \text{ N/m}^2 - (-29{,}430 \text{ N/m}^2)$$

$$P_1 = 176697.2 \text{ N/m}^2$$

$$\boxed{P_1 = 176.7 \text{ kPa}}$$

Problem 7.38

Arteries carry oxygen-rich blood from the heart through the body. The heart acts as a pump, causing the blood to flow through the arteries. Healthy arteries have smooth inner walls, and blood flows through them easily. However, arteries can 'clog' as a result of a buildup called plaque. This buildup can block blood flow, and make the heart have to work harder to supply blood to the body, which can eventually lead to a heart failure. The average healthy human femoral artery has a diameter of roughly 8 mm, and has an average volumetric flow rate 350 ml/min. If plaque buildup has reduced the effective diameter of an artery to only 2mm, what is the loss in blood pressure due to the clog?

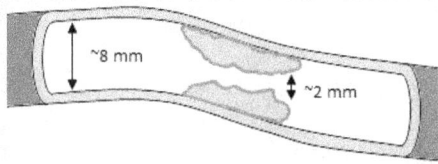

From table 7.3.1, human blood has approximate density of 1025 kg/m³.

The volumetric flow rate of blood through the femoral artery converted to SI units:
350 ml/min = 5.83 x 10⁻⁶ m/s

We start with the equation of continuity to find the fluid velocity through the clog:
$A_1 v_1 = A_2 v_2$
$\pi r_1^2 v_1 = \pi r_2^2 v_2$
$16 v_1 = v_2$

We find that the velocity of the blood must increase by 16 times through the clog. To find the drop in pressure, we use Bernoulli's equation (with no change in elevation), and then substitute in for v_2:

$$P_1 + \frac{1}{2}\rho v_1^2 + \cancel{\rho g h_1} = P_2 + \frac{1}{2}\rho v_2^2 + \cancel{\rho g h_2}$$

$$P_1 - P_2 = \Delta P = \frac{1}{2}\rho(16 v_1)^2 - \frac{1}{2}\rho v_1^2$$

$$\Delta P = \frac{1}{2}\rho(256 v_1^2 - v_1^2)$$

$$\Delta P = \frac{1}{2}\rho\, 255 v_1^2 = \frac{1}{2}(1025 \frac{kg}{m^3})(\,255 \cdot (5.83 \times 10^{-6} m/s)^2)$$

$$\Delta P = 4.44 \times 10^{-6}\ N/m^2$$

> $\Delta P = 4.44 \times 10^{-6}\ Pa$
> This may not seem like much, but consider the cumulative effect of all arteries and veins in the body with similar clogs!

Problem 7.39

A gate valve is a simple device used to control (or stop) the flow of a fluid through a pipe. A fully-open valve (100% open) provides no restriction to flow, and a fully closed valve (0% open) will completely shut off flow.

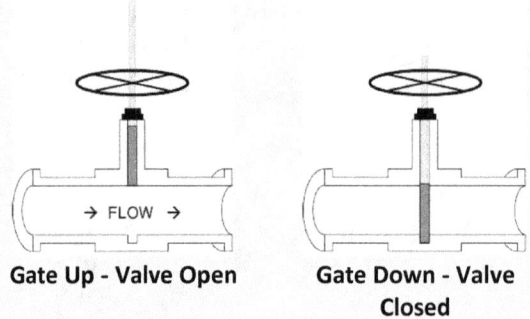

Gate Up - Valve Open **Gate Down - Valve Closed**

If you were to plot the outlet flow rate vs % open of the valve, would you expect the plot to be linear? Explain.

The plot will not be linear. The pressure remains constant, but the cross-sectional area does not increase linearly as the gate goes up and down (because the pipe is round).

Problem 7.40

A cylindrical tank is filled with 5 feet of water. At time 0, a plug is removed from the bottom and water starts draining through a 1 inch diameter hole. What is the initial velocity of the draining water? When the tank is half empty, what is the velocity of the draining water?

Fluid velocity at the exit is given by the following equation:
$$v = \sqrt{2\,g\,h}$$

At initial open, the height is 5 feet (1.524 meters):
$$v = \sqrt{2\left(9.8\,\frac{m}{s}\right)1.524\,m}$$
$$v = 5.47\,m/s$$

$$\boxed{v_{initial} = 5.47\,m/s}$$

When the tank is half full, the height is 2.5 feet (0.762 meters):
$$v = \sqrt{2\left(9.8\,\frac{m}{s}\right)0.762\,m}$$
$$v = 3.86\,m/s$$

$$\boxed{v_{half} = 3.86\,m/s}$$

Problem 7.41

Determine how long it will take for the system below to come to equilibrium once the valve between the two tanks once opened. Plot the height of both tanks against time to confirm your answer.

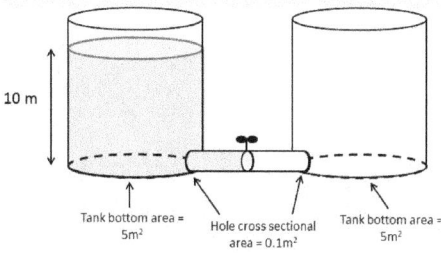

We must adapt the equation for velocity from a draining tank to account for the fact that the driving force is not just the pressure in tank 1, but rather the difference in pressures between the two tanks. This makes the equation simply:

$$v = \sqrt{2g(h_1 - h_2)}$$

We can now start to build our spreadsheet. In final form, it will look something like this (equations shown):

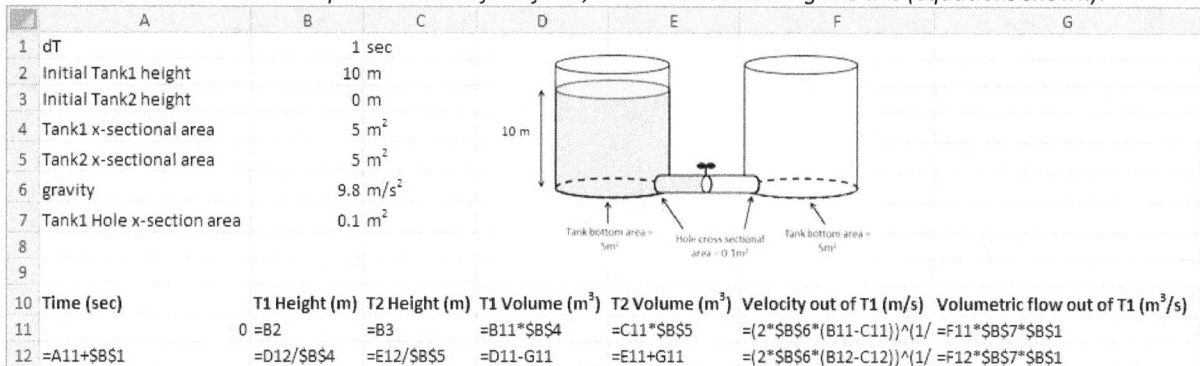

Dragging down to where the tanks reach the same height, we find a time of approximately **34 seconds**. A plot of the tank heights as a function of time confirms this:

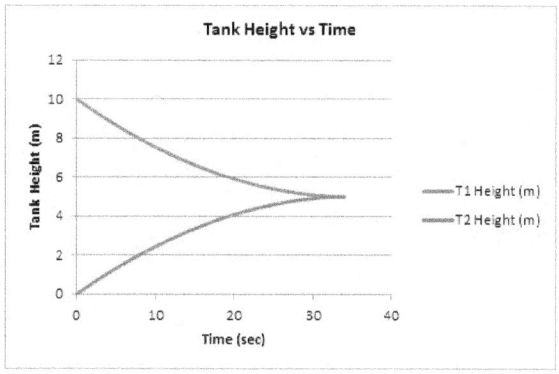

Problem 7.42

Your garden hose that is 1.7 cm in diameter, and the water is flowing through the hose at 3.2 m/s. If you use your thumb to cover the outlet and reduce the opening area by half, what is the exit velocity of the water?

This is a continuity equation problem:

$$A_1 v_1 = A_2 v_2$$

$$v_1 \pi \left(\frac{d_1}{2}\right)^2 = v_2 \pi \left(\frac{d_2}{2}\right)^2$$

The diameter is decreased by half, so set $d_2 = \frac{1}{2} d_1$:

$$v_1 \pi \left(\frac{d_1}{2}\right)^2 = v_2 \pi \left(\frac{d_1}{4}\right)^2$$

$$\frac{v_1 \pi \left(\frac{d_1}{2}\right)^2}{\pi \left(\frac{d_1}{4}\right)^2} = v_2$$

$$\frac{v_1 \frac{d_1^2}{4}}{\frac{d_1^2}{16}} = v_2$$

$$4 v_1 = v_2$$

$$4 \cdot 3.2 \text{ m/s} = v_2$$

$$12.8 \text{ m/s} = v_2$$

$$\boxed{v_2 = 12.8 \text{ m/s}}$$

Problem 7.43

10 m³/hr of water flows through a pipe with a 100 mm interior diameter. If the pipe goes through a constriction and is reduced to an 80 mm interior diameter, what is the new velocity?

We first find the velocity of the water in the larger section of the pipe:
$\dot{V} = A v \rightarrow v = \dot{V} / A$
v = (10 m³/hr) (hr / 3600 s) / (π (0.1 m/2)²)
v = (0.0028 m³/s) / (0.0079 m²)
v = 0.35 m/s

Next we find the velocity of the water in the smaller section of the pipe:
$\dot{V} = A v \rightarrow v = \dot{V} / A$
v = (10 m³/hr) (hr / 3600 s) / (π (0.04 m/2)²)
v = (0.0028 m³/s) / (0.0050 m²)
v = 0.56 m/s

$$\boxed{\begin{array}{l} v_{100} = 0.35 \text{ m/s} \\ v_{80} = 0.56 \text{ m/s} \end{array}}$$

Chapter 7 | Fluid Mechanics

Problem 7.44

Calculate the pressure drop when you cover the end of a hose with your thumb as in problem 41 of this chapter.

This is a Bernoulli equation problem. We start by cancelling the potential energy terms and consolidating the pressure and velocity terms.

$$P_{inlet} - P_{outlet} = \frac{1}{2}\rho(v_2^2 - v_1^2)$$

$$\Delta P = \frac{1}{2}\left(1000\ kg/m^3\right)\left((12.8\ m/s)^2 - (3.2\ m/s)^2\right)$$

$$\Delta P = \left(500\ kg/m^3\right)\left(163.84\ m^2/s^2 - 10.24\ m^2/s^2\right)$$

$$\Delta P = 76{,}800\ Pa$$

ΔP = 76.8 kPa

Problem 7.45

In reality, the density of air is in earth's atmosphere not constant and changes as a function of altitude. Given the incomplete table below, estimate as accurately as possible the atmospheric pressure at each elevation (pressure at sea level is of course known). Create a plot of air density and pressure as a function of altitude. Then, compare your estimated values to research on the web. How well do they match?

The pressure at a given elevation is simply due to the column of fluid above it. So, we will determine the pressure due to each individual segment, and then add up all of the segments above any given elevation. The first three lines of the table are shown below:

	A	B	C	D	E	F	G	H
1	Altitude (m)	Density of Air (kg/m³)	Atmospheric Pressure (kPa)		height range	Linear average of air density	Pressure from segment (Pa)	cumulative sum
2	0	1.225	101.325		=A3-A2	=(B2+B3)/2	=F2*9.8*E2	=SUM(G2:G$22)
3	500	1.167			=A4-A3	=(B3+B4)/2	=F3*9.8*E3	=SUM(G3:G$22)
4	1,000	1.112			=A5-A4	=(B4+B5)/2	=F4*9.8*E4	=SUM(G4:G$22)

However, once we complete the full table, we find that atmospheric pressure is not correct at sea level. This is because we are ignoring everything above 15,000 m – one unknown that we do not have is exactly where the earth's atmosphere ends (in other words, we don't have a definitive height for our column of fluid). To get our best estimate for the pressure due to that region, though, we can add in a 'correction' to each value that is equal to the difference between our calculated and the actual values:

	A	B	C	D	E	F	G	H	I	J	K
1	Altitude (m)	Density of Air (kg/m³)	Atmospheric Pressure (kPa)		height range	Linear average of air density	Pressure from segment (Pa)	cumulative sum	adjustment	estimated pressure (Pa)	estimated pressure (kPa)
2	0	1.225	101.325		=A3-A2	=(B2+B3)/2	=F2*9.8*E2	=SUM(G2:G$22)	=101325-$H$2	=H2+I2	=J2/1000
3	500	1.167			=A4-A3	=(B3+B4)/2	=F3*9.8*E3	=SUM(G3:G$22)	=101325-$H$2	=H3+I3	=J3/1000
4	1,000	1.112			=A5-A4	=(B4+B5)/2	=F4*9.8*E4	=SUM(G4:G$22)	=101325-$H$2	=H4+I4	=J4/1000

Our estimated values are shown below (according to the calculations above), and are shown alongside the actual measured atmospheric pressure values. You will notice they are relatively close to each other, which means we did an adequate job of our estimations. The plot of pressure vs altitude is also below.

Altitude (m)	Density of Air (kg/m³)	Atmospheric Pressure (kPa)	estimated pressure (kPa)	Actual Data
0	1.225	101.325	101.325	101.325
500	1.167		95.4646	95.46
1,000	1.112		89.88105	89.87
1,500	1.058		84.56455	84.55
2,000	1.006		79.50775	79.5
2,500	0.957		74.6984	74.7
3,000	0.909		70.1267	70.11
3,500	0.863		65.7853	65.87
4,000	0.819		61.6644	61.66
4,500	0.777		57.7542	57.75
5,000	0.736		54.04735	54.05
6,000	0.66		47.20695	47.22
7,000	0.59		41.08195	41.11
8,000	0.526		35.61355	35.66
9,000	0.467		30.74785	30.8
10,000	0.413		26.43585	26.5
11,000	0.365		22.62365	22.7
12,000	0.312		19.30635	19.4
13,000	0.266		16.47415	16.58
14,000	0.228		14.05355	14.17
15,000	0.195		11.98085	12.11

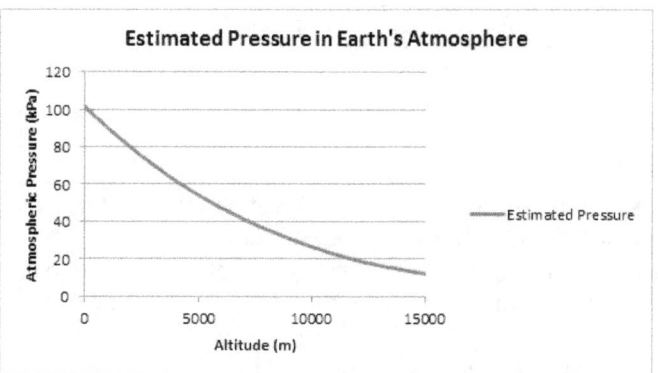

Chapter 7 | Fluid Mechanics

Problem 7.46

In reality, the density of air is in earth's atmosphere not constant and changes as a function of altitude. Given the incomplete table below, estimate as accurately as possible the atmospheric pressure at each elevation (pressure at sea level is of course known). Create a plot of air density and pressure as a function of altitude. Then, compare your estimated values to research on the web. How well do they match?

To solve this problem, we will take the following steps:
1. Calculate the measured volumetric flowrates
2. Plot the data points
3. Create a trendline and determine the equation of the relation between meter reading and flow rate
4. Use the equation to estimate the flow rates at the given readings.

We first use a spreadsheet to calculate the volumetric flow rate at each reading:

Equations and setup

	A	B	C	D
1	Flowmeter Reading	Collection Time (min)	Volume Collected (L)	Volumetric Flowrate (L/min)
2	1	2	0.871	0.4355
3	1	2	0.883	0.4415
4	2	2	1.342	0.671
5	2	2	1.324	0.662
6	3	1	0.88	0.88
7	3	1	0.874	0.874
8	4	1	1.093	1.093
9	4	1	1.111	1.111
10	5	1	1.3	1.3
11	5	1	1.339	1.339

Final table

	A	B	C	D
1	Flowmeter Reading	Collection Time (min)	Volume Collected (L)	Volumetric Flowrate (L/min)
2	1	2	0.871	=C2/B2
3	1	2	0.883	=C3/B3
4	2	2	1.342	=C4/B4
5	2	2	1.324	=C5/B5
6	3	1	0.88	=C6/B6
7	3	1	0.874	=C7/B7
8	4	1	1.093	=C8/B8
9	4	1	1.111	=C9/B9
10	5	1	1.3	=C10/B10
11	5	1	1.339	=C11/B11

We then create a scatter plot of the data. Once created, we click on the data series, and right click to bring up the options menu. Select "Add Trendline", which will bring up the 'Format Trendline' option box. To put the equation of the trendline on the chart, click the "Display Equation on chart" box.

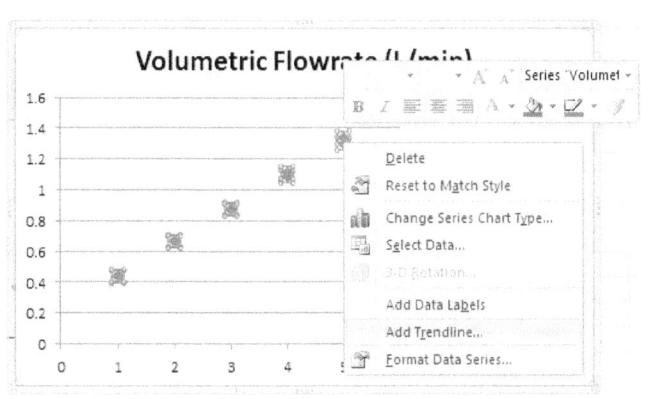
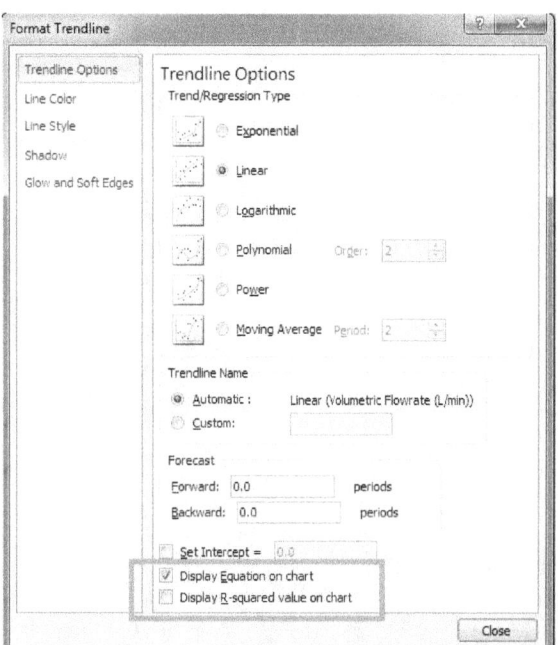

We now see a trendline on our chart along with an equation that gives us the relation of our flowmeter reading to the volumetric flowrate.

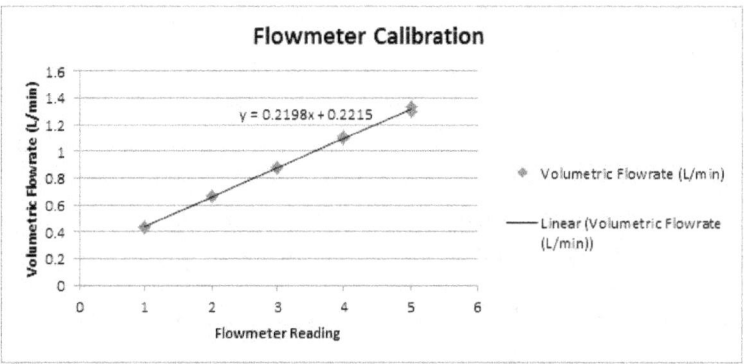

In this equation, Y is the volumetric flowrate and X is the flowmeter reading. We can then use the resulting equation to estimate the flowrates at the various flowmeter readings.
2.7 : 0.2198(2.7) + 0.2215 = **0.8150**
3.4 : 0.2198(3.4) + 0.2215 = **0.9689**
4.6 : 0.2198(4.6) + 0.2215 = **1.2326**

2.7 : ~0.8150 L/min
3.4 : ~0.9689 L/min
4.6 : ~1.2326 L/min

Chapter 8 Solutions
MATERIAL & ENERGY BALANCES

Problem 8.1

Extraction is a physical process in which a component of a mixture is selectively removed using a solvent that dissolves the desired component but does not mix with the undesired component. In the PFD below, component B is being extracted from a A-B mixture by component C.

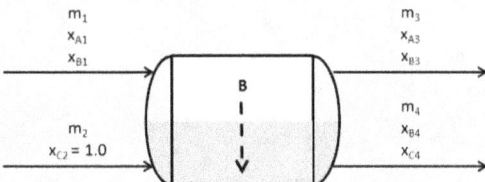

For the process shown, write the material balances for the following:
    a. Total mass
    b. Component A
    c. Component B
    d. Component C

a) Total mass balance:

$$\text{In} = \text{Out}$$
$$\boxed{m_1 + m_2 = m_3 + m_4}$$

b) Component A balance:

$$\text{In} = \text{Out}$$
$$\boxed{m_1 x_{A1} = m_3 x_{A3}}$$

c) Component B balance:

$$\text{In} = \text{Out}$$
$$\boxed{m_1 x_{A1} = m_3 x_{A3}}$$

d) Component C balance:

$$\text{In} = \text{Out}$$
$$m_2 x_{C2} = m_4 x_{C4} \text{ ; and } x_{C2} = 1, \text{ so:}$$
$$\boxed{m_2 = m_4 x_{C4}}$$

Problem 8.2

Reactant A and a small amount of catalyst C are fed to a reactor at steady-state. In the presence of the catalyst, A reacts to form component B. Write the material balances for the following:
   a. Total mass
   b. Component A
   c. Component B
   d. Component C

We begin by drawing a diagram of this process:

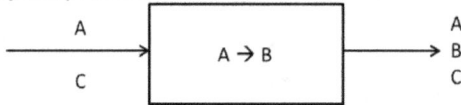

a) Total mass balance:
$$\text{In} = \text{Out}$$
$$\boxed{m_{in} = m_{out}}$$

b) Component A balance:
$$\text{In} - \text{Consumed} = \text{Out}$$
$$\boxed{m_{in}\, x_{A,in} - m_{A,consumed} = m_{out}\, x_{A,out}}$$

c) Component B balance (*B generates at a 1:1 ratio of A consumed*):
$$\text{Generated} = \text{Out}$$
$$\boxed{m_{A,consumed} = m_{out}\, x_{B,out}}$$

d) Component C balance (*C does not react!*):
$$\text{In} = \text{Out}$$
$$m_1\, x_{C1} = m_4\, x_{C4} \; ; \text{ and } x_{c2} = 1, \text{ so:}$$
$$\boxed{m_2 = m_4\, x_{C4}}$$

Chapter 8 | Material & Energy Balances

Problem 8.3

Two ethanol-water mixtures are contained in separate tanks. Tank 1 contains 40.0 wt% ethanol and tank 2 contains 70.0 wt% ethanol. If 200 kg from tank 1 is combined with 150 kg from tank 2, what is the composition of the final product?

This is a batch mixing process material balance. To find the final product composition we determine the amount of ethanol in the final product and divide by the mass of the final product.

Total mass balance:

$$\begin{aligned} \text{In} &= \text{Out} \\ 200 \text{ kg} + 150 \text{ kg} &= \text{Out} \\ 350 \text{ kg} &= \text{Out} \end{aligned}$$

Ethanol component mass balance:

$$\begin{aligned} \text{Ethanol in} &= \text{Ethanol out} \\ (0.40 \text{ wt\% ethanol})(200 \text{ kg}) + (0.70 \text{ wt\% ethanol})(150 \text{ kg}) &= \text{Ethanol out} \\ 80 \text{ kg ethanol} + 105 \text{ kg ethanol} &= \text{Ethanol out} \\ 185 \text{ kg ethanol} &= \text{Ethanol out} \end{aligned}$$

Composition % of ethanol:

$$\text{wt\% oil} = \frac{\text{mass ethanol}}{\text{total mass}} = \frac{185 \text{ kg ethanol}}{350 \text{ kg total}} = 52.9\,\%$$

> The final mixture is 350 kg total and is 52.9% ethanol by mass

Problem 8.4

1000 g of a 5.0 wt% NaCl-water solution is needed. You have a 20.0 wt% NaCl-water solution and pure water available. How much of each should be mixed?

This is a batch mixing process material balance. We will first need to determine the mass of NaCl in the final solution, then determine the mixing amounts. Let m_1 be the mass of the 20% NaCl solution, and m_2 the pure water.

NaCl component mass balance:

$$\begin{aligned} \text{NaCl in} &= \text{NaCl out} \\ m_1 (0.20 \text{ g NaCl/g solution}) &= 0.08\,(1000 \text{ g}) \\ m_1 (0.20 \text{ g NaCl/g solution}) &= 80 \text{ g NaCL} \\ m_1 &= 400 \text{ g solution} \end{aligned}$$

Total mass balance:

$$\begin{aligned} \text{In} &= \text{Out} \\ 400 \text{ g} + m_2 &= 1000 \text{ g} \\ m_2 &= 600 \text{ g} \end{aligned}$$

> We need to mix 400 g of the 20% NaCl solution, and 600 g of pure water

Problem 8.5

A 50 wt% methanol-water mixture is fed at steady-state to a separator at 300 kg/hr, and leaves in two fractions. The mass flow rate of methanol in the top stream is 135 kg/hr, and the mass flow rate of water in the bottom stream is 145 kg/hr. What are the total mass flows and compositions of both streams?

We begin by drawing a diagram of this process:

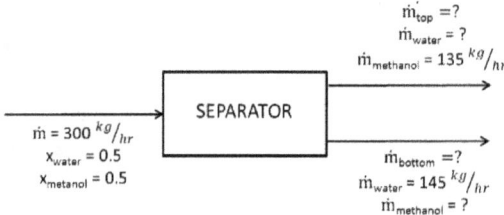

Methanol mass balance:

	In	=	Out
	300 kg/hr (0.5 kg methanol/kg)	=	135 kg methanol/hr + $m_{methanol}$
	150 kg/hr methanol	=	135 kg methanol/hr + $m_{methanol}$
	15 kg/hr methanol	=	$m_{methanol}$

Water mass balance:

	In	=	Out
	300 kg/hr (0.5 kg water/kg)	=	145 kg methanol/hr + m_{water}
	150 kg/hr water	=	145 kg methanol/hr + m_{water}
	5 kg/hr water	=	m_{water}

Top stream:
m_{top} = 135 kg/hr methanol + 5 kg/hr water
m_{top} = *140 kg/hr*

Bottom stream:
m_{bottom} = 145 kg/hr water + 15 kg/hr methanol
m_{bottom} = *160 kg/hr*

$$\dot{m}_{top} = 140 \text{ kg/hr}$$
$$\dot{m}_{bottom} = 160 \text{ kg/hr}$$
$$\dot{m}_{methanol} = 15 \text{ kg/hr methanol}$$
$$\dot{m}_{water} = 5 \text{ kg/hr water}$$

Problem 8.6

Air is humidified by evaporating liquid water into an air stream. The desired product contains 15.0 mol% water. Air is fed to a humidifier at steady-state at a rate of 10 mol/hr. What is the required flow of water?

We begin by drawing a diagram of this process:

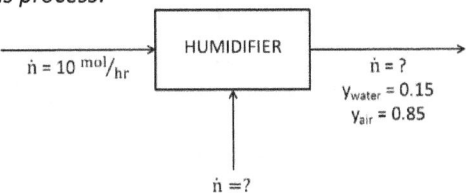

There is no reaction, so there is no generation/consumption. Total mole balance:

$$\text{In} + \cancel{\text{Generated}} = \text{Out} + \cancel{\text{Consumed}}$$

$$10 \text{ mol/hr} + n_{water} = n_{out}$$

$$n_{water} = n_{out} - 10 \text{ mol/hr}$$

Air mole balance:

$$\text{Air in} = \text{Air out}$$

$$10 \text{ mol/hr} = n_{out}(0.85)$$

$$11.764 \text{ mol/hr} = n_{out}$$

We now have enough information to solve for the required flow of water:

$$n_{water} = n_{out} - 10 \text{ mol/hr}$$

$$n_{water} = 11.764 \text{ mol/hr} - 10 \text{ mol/hr}$$

$$\mathbf{n_{water} = 1.764 \text{ mol/hr}}$$

The required flow of water is 1.764 mol/hr

Problem 8.7

Oxygen is separated from the air in a process called cryogenic distillation. A cryo unit has a feed of 500 mol/hr dry air (79 mol% N_2 and 21 mol% O_2) as shown in the PFD below. Solve for the process unknowns ($\dot{n}_2, \dot{n}_{2,O2}, \dot{n}_3, \dot{n}_{2,N2}$).

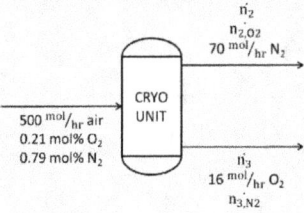

Total mole balance: $500 = n_2 + n_3$

O_2 component balance:

$$
\begin{aligned}
O_2 \text{ in} &= O_2 \text{ out} \\
(500 \text{ mol/hr})(0.21 \text{ mol\% } O_2) &= n_{2,O2} + 16 \text{ mol/hr } O_2 \\
105 \text{ mol/hr } O_2 &= n_{2,O2} + 16 \text{ mol/hr } O_2 \\
89 \text{ mol/hr } O_2 &= n_{2,O2}
\end{aligned}
$$

N_2 component balance:

$$
\begin{aligned}
N_2 \text{ in} &= N_2 \text{ out} \\
(500 \text{ mol/hr})(0.79 \text{ mol\% } N_2) &= 70 \text{ mol/hr } N_2 + n_{3,N2} \\
395 \text{ mol/hr } N_2 &= 70 \text{ mol/hr } N_2 + n_{3,N2} \\
325 \text{ mol/hr } N_2 &= n_{3,N2}
\end{aligned}
$$

Stream 2:
$n_2 = 89 \text{ mol/hr } O_2 + 70 \text{ mol/hr } N_2$
$n_2 = 159 \text{ mol/hr}$

Stream 3:
$n_3 = 16 \text{ mol/hr } O_2 + 325 \text{ mol/hr } N_2$
$n_3 = 341 \text{ mol/hr}$

$$
\boxed{\begin{aligned}
\dot{n}_2 &= 159 \text{ mol/hr} \\
\dot{n}_{2,O2} &= 89 \text{ mol/hr } O_2 \\
\dot{n}_3 &= 341 \text{ mol/hr} \\
\dot{n}_{2,N2} &= 325 \text{ mol/hr } N_2
\end{aligned}}
$$

Problem 8.8

A 1:1 mass ratio mixture of ethanol and water is fed to a distillation column at 1000 kg/hr. The bottoms stream of unknown composition has a mass flow rate of 600 kg/hr. The top stream has a composition of 96.0 wt% ethanol. Solve for the flow rate of the top stream and the composition of the bottom stream.

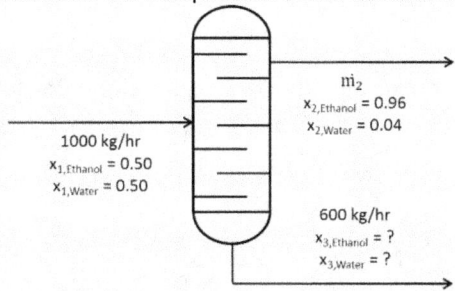

Total mass balance:

$$
\begin{aligned}
\text{In} &= \text{Out} \\
1000 \text{ kg/hr} &= \dot{m}_2 + 600 \text{ kg/hr} \\
400 \text{ kg/hr} &= \dot{m}_2
\end{aligned}
$$

Ethanol component balance:

$$
\begin{aligned}
\text{Ethanol in} &= \text{Ethanol out} \\
(1000 \text{ kg/hr})(0.50 \text{ wt\% ethanol}) &= 0.96\,(\dot{m}_2) + 600 \text{ kg/hr}\,(x_{3,\text{ethanol}}) \\
500 \text{ kg/hr ethanol} &= 0.96\,(400 \text{ kg/hr}) + 600 \text{ kg/hr}\,(x_{3,\text{ethanol}}) \\
500 \text{ kg/hr ethanol} &= 384 \text{ kg/hr ethanol} + 600 \text{ kg/hr}\,(x_{3,\text{ethanol}}) \\
116 \text{ kg/hr ethanol} &= 600 \text{ kg/hr}\,(x_{3,\text{ethanol}}) \\
0.193 &= x_{3,\text{ethanol}}
\end{aligned}
$$

Water component balance:

$$
\begin{aligned}
\text{Water in} &= \text{Water out} \\
(1000 \text{ kg/hr})(0.50 \text{ wt\% water}) &= 0.04\,(\dot{m}_2) + 600 \text{ kg/hr}\,(x_{3,\text{water}}) \\
500 \text{ kg/hr water} &= 0.04\,(400 \text{ kg/hr}) + 600 \text{ kg/hr}\,(x_{3,\text{water}}) \\
500 \text{ kg/hr water} &= 16 \text{ kg/hr water} + 600 \text{ kg/hr}\,(x_{3,\text{water}}) \\
484 \text{ kg/hr water} &= 600 \text{ kg/hr}\,(x_{3,\text{water}}) \\
0.806 &= x_{3,\text{water}}
\end{aligned}
$$

$$
\boxed{\begin{aligned}
\dot{m}_2 &= 400 \text{ kg/hr} \\
x_{3,\text{water}} &= 0.193 \\
x_{3,\text{ethanol}} &= 0.806
\end{aligned}}
$$

Problem 8.9

Skim milk is made by removing most of the fat from whole milk. Whole milk containing 4.5% fat is fed to a separator. The final skim milk product contains 90.5% water, 3.5% proteins, 5.1% carbohydrates, 0.1% fat and 0.8% solids. Determine the full composition of the feed milk assuming the only mass removed was the fat.

Start by considering a basis of 100g of skim milk output. This lets us set up both the total mass balance and the fat component balance. Let X = the mass of the whole milk, and Y = the mass of the fats removed.

Total mass balance:

$$\text{In} = \text{Out}$$
$$X = 100 + Y$$
$$\mathbf{X - 100 = Y}$$

Fat component balance:

$$\text{In} = \text{Out}$$
$$(0.045)X = (0.001)(100) + Y$$
$$0.045X = 0.1 + Y$$

next we substitute in for Y using the equation from the total mass balance:

$$0.045X = 0.1 + Y \quad \leftarrow Y = X - 100$$
$$0.045X = 0.1 + X - 100$$
$$99.9 = 0.955X$$
$$\mathbf{104.6 = X}$$

We can now do each individual component balance, solving for the initial % compositions:

$X_{carb}(104.6g) = (100g)(0.035)$
$X_{carb} = 3.5g/104.6g = 0.033$
$\mathbf{X_{carb} = 3.3\%}$

$X_{protein}(104.6g) = (100g)(0.051)$
$X_{protein} = 5.1g/104.6g = 0.048$
$\mathbf{X_{protein} = 4.8\%}$

$X_{water}(104.6g) = (100g)(0.905)$
$X_{water} = 90.5g/104.6g = 0.865$
$\mathbf{X_{water} = 86.5\%}$

$X_{carb}(104.6g) = (100g)(0.008)$
$X_{carb} = 0.8g/104.6g = 0.0076$
$\mathbf{X_{carb} = 0.76\%}$

Whole Milk Composition:
4.5% fat
3.3% protein
4.8% carbohydrates
86.5% water
0.76% solids

Problem 8.10

Hydrogen sulfide (H_2S) is a gas commonly found dissolved in ground water. It imparts a "rotten egg" odor and acts as a weak acid, which can corrode piping and turn water black. A common method of removing this contaminant is to bubble air through the water, allowing the H_2S gas to escape into the air. An aeration process is shown in the BFD below. Solve for the mass flow rate of air.

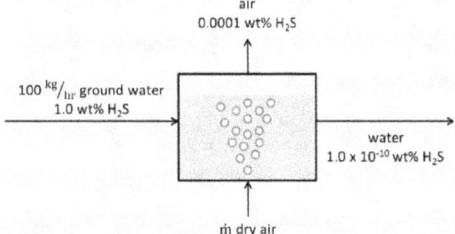

We start with a component balance around the H_2S:

$$H_2S \text{ in} = H_2S \text{ out}$$
$$(100 \text{ kg/hr water})(0.01 \text{ kg } H_2S/\text{kg water}) = 0.0001 (m_{air,out}) + 1.0 \times 10^{-10} (m_{water,out})$$
$$1 \text{ kg/hr } H_2S = 0.0001 (m_{air,out}) + 1.0 \times 10^{-10} (m_{water,out})$$

Because the mass of H_2S is so small at 0.1 ppb, we can treat the water out stream as pure water:

$$\text{Water in} = \text{Water out}$$
$$(1000 \text{ kg/hr})(0.99 \text{ wt\% water}) = 0.9999999999 (m_{water,out})$$
$$99 \text{ kg/hr water} = \cancel{0.9999999999} \approx 1 \cdot (m_{water,out})$$
$$99 \text{ kg/hr water} = (m_{water,out})$$

Now plugging the value for $m_{water,out}$ back into our H_2S balance:

$$1 \text{ kg/hr } H_2S = 0.0001 (m_{air,out}) + 1.0 \times 10^{-10} (m_{water,out})$$
$$1 \text{ kg/hr } H_2S = 0.0001 (m_{air,out}) + 1.0 \times 10^{-10} (99 \text{ kg/hr water})$$
$$1 \text{ kg/hr } H_2S = 0.0001 (m_{air,out}) + 9.9 \times 10^{-9} \text{ kg/hr } H_2S$$
$$0.9999999901 \text{ kg/hr } H_2S = 0.0001 (m_{air,out})$$
$$9999.999901 \text{ kg/hr} = m_{air,out}$$
$$10,000 \text{ kg/hr} \approx m_{air,out}$$

$$\boxed{m_{air,in} \approx 10,000 \text{ kg/hr}}$$

Problem 8.11

Jam is made by mixing crushed berries and sugar in a 1:1 mass ratio, and then heating the mixture to evaporate off water until the reduction is only 30% water by mass. How many kilograms of jam can be made with one kilogram of berries?

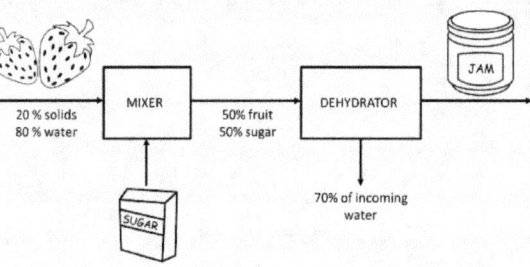

Begin with the total mass balance:

$$\begin{aligned} \text{In} &= \text{Out} \\ \text{Berries} + \text{Sugar} &= \text{Water}_{evap} + \text{Jam} \\ 1 \text{ kg} + 1 \text{ kg} &= \text{Water}_{evap} + \text{Jam} \\ 2 \text{ kg} &= \text{Water}_{evap} + \text{Jam} \end{aligned}$$

This gives the equation: Jam = 2 kg - Water$_{evap}$

Now doing a water component balance:

$$\begin{aligned} \text{Water in} &= \text{Water out} \\ (1 \text{ kg berries})(0.80 \text{ kg water/kg berries}) &= 0.70 \,(\text{Water}_{evap}) \\ 0.8 \text{ kg water} &= 0.70 \,(\text{Water}_{evap}) \\ 1.14 \text{ kg} &= \text{Water}_{evap} \end{aligned}$$

Plugging into the jam equation:

Jam = 2 kg - Water$_{evap}$

Jam = 2 kg - *1.14 kg*

Jam = 0.86 kg

We can make 0.86 kg of Jam

Problem 8.12
Distilled drinking water is made from a water source with a total dissolved solids content of 0.07 mg/L. The water is first put through a course filter which removes 20% of the solids. 50% of the water is then distilled off as pure water and sent to be bottled. The rest of the water is then returned back to the water source. What is the solids content (in mg/L) of the return water?

We begin by drawing a diagram of this process:

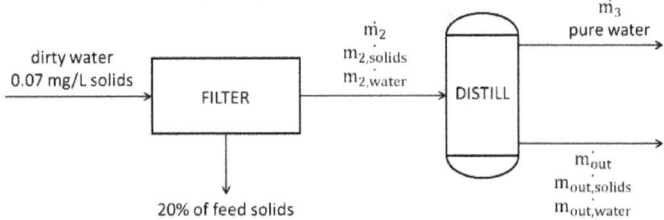

We will use a basis of 1000 L dirty water fed to the process.
1000 L dirty water = 1 kg water

Next determine the amount of solids in the feed water:

0.07 mg solids	1000 L water	1×10^{-6} kg	= 7.0×10^{-5} kg solids
L water		mg	

Overall solids balance:

$$\begin{aligned}
\text{Solids in} &= \text{Solids out} \\
7.0 \times 10^{-5} \text{ kg solids} &= 0.20 \text{ (feed)} + m_{out,solids} \\
7.0 \times 10^{-5} \text{ kg solids} &= 0.20 \,(7.0 \times 10^{-5} \text{ kg solids}) + m_{out,solids} \\
5.6 \times 10^{-5} \text{ kg solids} &= m_{out,solids}
\end{aligned}$$

Overall water balance:

$$\begin{aligned}
\text{Water in} &= \text{Water out} \\
1000 \text{ kg} &= \dot{m}_3 + m_{out,water} \\
1000 \text{ kg} &= 0.50 \text{ (feed)} + m_{out,water} \\
1000 \text{ kg} &= 500 \text{ kg} + m_{out,water} \\
500 \text{ kg} &= m_{out,water}
\end{aligned}$$

Final solids concentration:

5.6×10^{-5} kg solids	mg	1 kg water	= 0.112 mg/L solids
500 kg water	1×10^{-6} kg	1 L water	

The return water is 0.112 mg/L solids

Problem 8.13

A CSTR has an inlet flow of 200 L/min and an average residence time of 4 hours. How large is the reactor?

$$\text{Residence time} = \frac{\text{system capacity to hold a substance}}{\text{flow rate of substance through system}}$$

$$4 \text{ hr} = \frac{V}{200 \text{ L/min}}$$

$$(4 \text{ hr})(200 \text{ L/min})(60 \text{ min/hr}) = V$$

$$48{,}000 \text{ L} = V$$

Tank volume is 48,000 L

Problem 8.14

A 180-lb man with estimated blood volume of 6.12 L is being administered a drug intravenously. The intent is to maintain drug concentration in the blood at 200 µg/L. Consumption of the drug due to fighting infection is 35 µg/day, and drug excretion is 50 µg/day. Determine the residence time of the drug in the man's body.

$$\text{Residence time} = \frac{(6.12 \text{ L})(200 \frac{\mu g}{L})}{35 \frac{\mu g}{day} + 50 \frac{\mu g}{day}}$$

$$\text{Residence time} = \frac{1224 \text{ } \mu g}{85 \frac{\mu g}{day}}$$

$$\text{Residence time} = \frac{1224 \text{ } \mu g}{85 \frac{\mu g}{day}}$$

$$\text{Residence time} = 14.4 \text{ } days$$

Residence time in body is 14.4 days

Problem 8.15

Component A is fed at 10 mol/min to a steady-state CSTR, where it reacts at a rate of 0.8 mol A/min·gal. The product stream contains only 15% of the original feed. What is the volume of the CSTR?

The molar flow out (F_A) is only 15% of the inlet (F_{A0}):
$F_A = 0.15 \cdot F_{A0}$
$F_A = 0.15 \cdot (10 \text{ mol/min})$
$F_A = 1.5 \text{ mol/min}$

The reactor volume can be found by:

$$V = \frac{F_{A0} - F_A}{-r_A}$$

$$V = \frac{10 \frac{\text{mol}}{\text{min}} - 1.5 \text{ mol/min}}{-0.8 \text{ mol/min} \cdot \text{gal}}$$

$$V = \frac{8.5 \text{ mol/min}}{-0.8 \text{ mol/min} \cdot \text{gal}}$$

$$V = 10.625 \text{ gal}$$

Reactor volume is 10.625 gallons

Problem 8.16

You are a process engineer in charge of a 10,000 gallon CSTR at a chemical plant. The reactor has a steady-state feed of 500 gallons/minute and a conversion of 70%. Your boss wants to double the product, so he instructs you to double your feed rate. This will double your raw material cost, but your boss believes that this action will double that product and thus is a smart move. Prove to your boss that his assumption is mistaken.

We start by comparing how the residence time of the reactor will change under the new feed rate:

$$\tau_{initial} = \frac{V}{\dot{V}}$$
$$\tau_{initial} = \frac{10,000 \text{ gal}}{500 \text{ gal/min}}$$
$$\tau_{initial} = 20 \text{ min}$$

$$\tau_{final} = \frac{V}{\dot{V}}$$
$$\tau_{final} = \frac{10,000 \text{ gal}}{1,000 \text{ gal/min}}$$
$$\tau_{final} = 10 \text{ min}$$

The reaction rate of the process is found by:

$$C = C_0 \, e^{-k\tau}$$
$$(0.3) = (1) \, e^{-k(20 \text{ min})}$$
$$\ln(0.3) = -k \, (20 \text{ min})$$
$$-1.204 = -k \, (20 \text{ min})$$
$$0.06 = k \text{ min}^{-1}$$

We then show that with a lower residence time, the final concentration (the output) will be less than 70%:

$$C = C_0 \, e^{-k\tau}$$
$$C = (1) \, e^{-0.06 \text{ min}^{-1}(20 \text{ min})}$$
$$C = e^{-0.6}$$
$$C = 0.549$$

Doubling the feed rate will decrease the overall conversion to 54.9%

Problem 8.17

5 mol/min of methane and 10 mol/min of dry air (79 mol% N_2 and 21 mol% O_2) are fed to a combustion chamber at steady-state. The combustion of methane occurs according to the balanced reaction:

$$CH_4 + 2\,O_2 \rightarrow CO_2 + 2\,H_2O$$

What is the composition of gas leaving the chamber? *Hint: The O_2 in air is a limiting reactant*

We begin by drawing a diagram of this process:

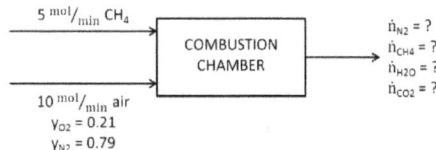

Methane and O_2 react in a 1:2 ratio. There are 5 mol/min of methane, but only 2.1 mol/min of O_2. O_2 is a limiting reactant. Determine the amount of CH_4 consumed:

$$2.1\,\frac{\text{mol}}{\text{min}\,O_2} \cdot \frac{1\,\text{mol}\,CH_4}{2\,\text{mol}\,O_2} = 1.05\,\text{mol}\,CH_4\,\text{consumed}$$

Methane component balance:

$$\begin{aligned}
\text{In} - \text{Consumed} &= \text{Out} \\
5\,\text{mol/min} - 1.05\,\text{mol/min} &= n_{CH4} \\
3.85\,\text{mol/min} &= n_{CH4}
\end{aligned}$$

Nitrogen component balance:

$$\begin{aligned}
\text{In} &= \text{Out} \\
0.79\,(10\,\text{mol/min}) &= n_{N2} \\
7.9\,\text{mol/min} &= n_{N2}
\end{aligned}$$

Water component balance:

$$\begin{aligned}
\text{Produced} &= \text{Out} \\
2.1\,\frac{\text{mol}}{\text{min}\,O_2} \cdot \frac{2\,\text{mol}\,H_2O}{2\,\text{mol}\,O_2} &= n_{H2O} \\
2.1\,\text{mol/min} &= N_{H2O}
\end{aligned}$$

Carbon dioxide component balance:

$$\begin{aligned}
\text{Produced} &= \text{Out} \\
2.1\,\frac{\text{mol}}{\text{min}\,O_2} \cdot \frac{1\,\text{mol}\,CO_2}{2\,\text{mol}\,O_2} &= n_{CO2} \\
1.05\,\text{mol/min} &= n_{CO2}
\end{aligned}$$

$$\boxed{\begin{aligned}
\dot{n}_{N2} &= 7.9\,\text{mol/min}\,N_2 \\
\dot{n}_{CH4} &= 3.85\,\text{mol/min}\,CH_4 \\
\dot{n}_{H2O} &= 2.1\,\text{mol/min}\,H_2O \\
\dot{n}_{CO2} &= 1.05\,\text{mol/min}\,CO_2
\end{aligned}}$$

Problem 8.18

Hydrogen peroxide decomposes into oxygen and water according to the balanced reaction:
$$2\,H_2O_2 \rightarrow O_2 + 2\,H_2O$$
If 100 kg/hr of hydrogen peroxide is fed to a reactor at steady-state where 80% of it decomposes, determine the resulting mass fractions of oxygen and water in the product stream.

We begin by drawing a diagram of the process:

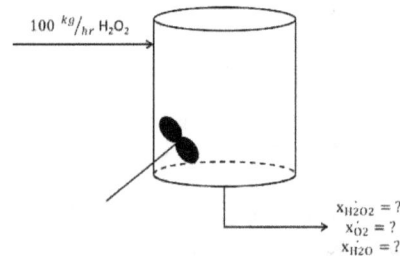

Because this is a reactive process we need to deal with moles of material:

$$\frac{100 \text{ kg } H_2O_2}{\text{hr}} \cdot \frac{\text{kmol}}{34 \text{ kg}} = 2.94 \text{ kmol/hr } H_2O_2$$

80% of the H_2O_2 reacts, so the hydrogen peroxide component balance is:

$$H_2O_2 \text{ In} = H_2O_2 \text{ Out} + H_2O_2 \text{ Consumed}$$
$$H_2O_2 \text{ In} = H_2O_2 \text{ Out} + 0.8\,(H_2O_2 \text{ In})$$
$$0.2\,(H_2O_2 \text{ In}) = H_2O_2 \text{ Out}$$
$$0.2\,(2.94 \text{ kmol/hr } H_2O_2) = H_2O_2 \text{ Out}$$
$$0.588 \text{ kmol/hr } H_2O_2 = H_2O_2 \text{ Out}$$

Oxygen component balance:

$$O_2 \text{ Generated} = O_2 \text{ Out}$$
$$(1 \text{ mol } O_2 \text{ Generated}/ 2 \text{ mol } H_2O_2 \text{ Consumed})(0.8 \cdot 2.94 \text{ kmol/hr } H_2O_2) = O_2 \text{ Out}$$
$$1.176 \text{ kmol } O_2/\text{hr} = O_2 \text{ Out}$$

Water component balance:

$$H_2O \text{ Generated} = H_2O \text{ Out}$$
$$(1 \text{ mol } H_2O \text{ Generated}/ 2 \text{ mol } H_2O_2 \text{ Consumed})(0.8 \cdot 2.94 \text{ kmol/hr } H_2O_2) = H_2O \text{ Out}$$
$$2.368 \text{ kmol } H_2O/\text{hr} = H_2O \text{ Out}$$

Component	mol product	molar mass	mass product	mass fraction
H_2O_2	0.588 kmol/hr	34 kg/kmol	19.992 kg	0.321
O_2	1.176 kmol/hr	32 kg/kmol	37.632 kg	0.603
H_2O	2.368 kmol	2 kg/kmol	4.736 kg	0.076
			62.360 kg	1.000

$$x_{H2O} = 0.603,\ x_{H2O2} = 0.076$$

Problem 8.19

Bioethanol fermentation occurs as a batch process. A 50 L liquid batch has an original gravity of 1.045. After fermentation is complete the final gravity of the liquid is 1.014.

    a. How much CO_2 was released during fermentation?
    b. How much ethanol was produced?

Divide the change in density by the molar mass of CO_2 to determine the amount of CO_2 produced per unit volume:

$$\frac{\Delta \text{ density (g/cm}^3)}{\text{molar mass CO}_2 \left(\frac{g}{mol}\right)} = \frac{\text{mol CO}_2}{\text{cm}^3}$$

$$\frac{1.045 \frac{g}{cm^3} - 1.014 \frac{g}{cm^3}}{44 \left(\frac{g}{mol}\right)} = \frac{\text{mol CO}_2}{\text{cm}^3}$$

$$7.045 \times 10^{-4} = \frac{\text{mol CO}_2}{\text{cm}^3}$$

Since the molar ratio of carbon dioxide to ethanol is 1:1:

$$7.045 \times 10^{-4} = \frac{\text{mol ethanol}}{\text{cm}^3}$$

a) Mass of CO_2 produced:

$$\frac{7.045 \times 10^{-4} \text{ mol CO}_2}{cm^3} \left|\frac{1000 \text{ cm}^3}{1 \text{ L}}\right| 50 \text{ L} \left|\frac{44 \text{ g CO}_2}{\text{mol CO}_2}\right. = 1549.9 \text{ g CO}_2$$

$$\boxed{1.5499 \text{ kg CO}_2}$$

b) Mass of ethanol produced:

$$\frac{7.045 \times 10^{-4} \text{ mol ethanol}}{cm^3} \left|\frac{1000 \text{ cm}^3}{1 \text{ L}}\right| 50 \text{ L} \left|\frac{46.07 \text{ g ethanol}}{\text{mol CO}_2}\right. = 1622.8 \text{ g CO}_2$$

$$\boxed{1.6228 \text{ kg ethanol}}$$

Problem 8.20

A common method of increasing overall product yields in chemical processes is to utilize a *recycle stream*, in which a portion of the downstream product is fed back to the reactor. Consider the BFD below for a reactor that has only 50% conversion of component A into B. By splitting the product stream into two equal streams and returning one to the reactor, the overall process exhibits more than 50% conversion of A. Solve for the composition of the final product stream, showing it is >50% component B.

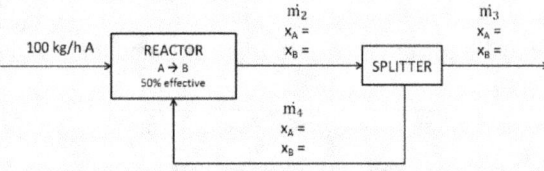

Total overall mass balance:
$$\text{In} = \text{Out}$$
$$100 \text{ kg/hr} = m_3$$

The splitter results in two equal streams, so $m_3 = m_3$:
$$100 \text{ kg/hr} = m_4$$

Mass balance over splitter:
$$\text{In} = \text{Out}$$
$$m_2 = m_3 + m_4$$
$$m_2 = 100 \text{ kg/hr} + 100 \text{ kg/hr}$$
$$m_2 = 200 \text{ kg/hr}$$

Component A balance over reactor:
$$\text{A in} - \text{A consumed} = \text{A out}$$
$$m_1 + m_4\, x_{4,A} - 0.5(m_1 + m_4\, x_{4,A}) = m_2\, x_{2,A}$$
$$100 \text{ kg/hr} + 100\, x_{4,A} - 0.5(100 \text{ kg/hr} + 100\, x_{4,A}) = 200\, x_{2,A}$$
$$50 \text{ kg/hr} + 50\, x_{4,A} = 200\, x_{2,A}$$

Because the compositions of m_2, m_3, and m_4 are the same, $x_{4,A} = x_{2,A}$:
$$50 \text{ kg/hr} + 50\, x_{2,A} = 200\, x_{2,A}$$
$$50 \text{ kg/hr} = 150\, x_{2,A}$$
$$0.33 = x_{2,A}$$

Product stream composition:
$x_{3,A} = 0.33$
$x_{3,B} = 1 - x_{3,A} = 0.67$

> $x_{3,A} = 0.33$
> $x_{3,B} = 0.67$
> >50% conversion!

Problem 8.21

Water at 25 °C is fed to a combustion furnace at a rate of 1000 kg/hr, where it is turned into steam and leaves at 120 °C. At what rate is energy being fed to the furnace?

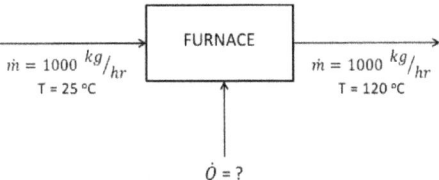

The water being fed to the reactor must be raised to 100 °C, vaporized, and then then raised (as steam) to 120 °C. Our energy flow rate:

$\dot{Q} = \dot{m}\Delta Tc_{water} + \dot{m}L_v + \dot{m}\Delta Tc_{steam}$

$\dot{Q} = \dot{m}[\ \Delta Tc_{water} + L_v + \Delta Tc_{steam}\ ]$

$\dot{Q} = 1000\ ^{kg}/_{hr}[\ (100°C - 25\ °C)(4.184\ ^{kJ}/_{kgC}) + 2260^{kJ}/_{kg} + (120°C - 100\ °C)(2.01\ k^{kJ}/_{kgC})\ ]$

$\dot{Q} = 1000\ ^{kg}/_{hr}[\ 313.8^{kJ}/_{kg} + 2260\ ^{kJ}/_{kg} + 40.2^{kJ}/_{kgC}\ ]$

$\dot{Q} = 2.614 \times 10^6\ kJ$

$\boxed{\dot{Q} = 2.614 \times 10^6\ kJ}$

Problem 8.22

Your objective is to design a heat exchanger that will get two streams to approximately the same exit temperature. Should you use co-current or counter-current flow?

Co-current flow would be the best choice. In co-current, as the fluids travel the length of the heat exchanger the difference in temperature between them decreases.

Problem 8.23

Hot process air is cooled by flowing through a counter-flow heat exchanger cooled by ethylene glycol. Using the BFD below, determine the exit temp of the process air. *Hint: Use appendix C.5.*

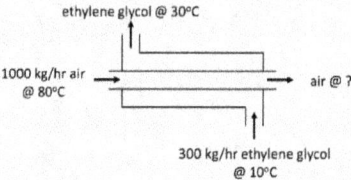

Appendix C5 - the heat capacity of ethylene glycol is 2.408 kJ/kg°C and the heat capacity of air is 1.0076 kJ/kg°C.

$$Q_{Lost} = Q_{Gained}$$
$$m_{air} \Delta T \, c_{air} = m_{glycol} \Delta T \, c_{glycol}$$
$$(1000 \text{ kg/hr})(80\,°C - T)(1.0076 \text{ kJ/kg °C}) = (300 \text{ kg/hr})(10\,°C - 30\,°C)(2.408 \text{ kJ/kg °C})$$
$$1007.6 \text{ kJ/hr°C } (80\,°C - T) = 14{,}448 \text{ kJ/hr}$$
$$80\,°C - T = 14.339\,°C$$
$$-T = -65.66\,°C$$

Air exit temp = 65.66 °C

Problem 8.24

Two water streams of different mass flowrates and different temperatures mix in a continuous steady-state process, and energy is added to the mixer in increase the outlet temperature. Given the mixing process below, determine the unknowns ($\dot{m}_{in}$, $\dot{m}_{out}$).

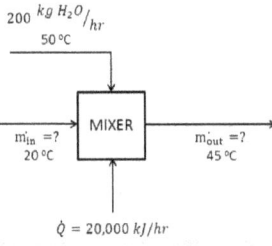

For simplicity we will set our reference point at 0°C.

Overall energy balance:

$$\text{In} = \text{Out}$$

$\dot{m}_{in}(20\ °C)(4.184\ kJ/kg\ °C) + 200\ kg/hr\ (50\ °C)(4.184\ kJ/kg\ °C) + 20{,}000\ kJ/hr = \dot{m}_{out}\ (45\ °C)\ (4.184\ kJ/kg\ °C)$

$\dot{m}_{in}(83.68\ kJ/kg) + 41{,}840\ kJ/hr + 20{,}000\ kJ/hr = \dot{m}_{out}\ (188.28\ kJ/kg)$

$\dot{m}_{in}(83.68\ kJ/kg) + 61{,}840\ kJ/hr = \dot{m}_{out}\ (188.28\ kJ/kg)$

Now us the overall mass balance to get one variable in terms of the other:

$$\text{In} = \text{Out}$$
$$\dot{m}_{in} + 200\ kg/hr = \dot{m}_{out}$$

and plug back into the energy balance:

$\dot{m}_{in}(83.68\ kJ/kg) + 61{,}840\ kJ/hr = \dot{m}_{out}\ (188.28\ kJ/kg)$
$\dot{m}_{in}(83.68\ kJ/kg) + 61{,}840\ kJ/hr = (\dot{m}_{in} + 200\ kg/hr)\ (188.28\ kJ/kg)$
$\dot{m}_{in}(83.68\ kJ/kg) + 61{,}840\ kJ/hr = 188.28\ kJ/kg\ \dot{m}_{in} + 37{,}656\ kJ/hr$
$24{,}184\ kJ/hr = 104.6\ kJ/kg\ \dot{m}_{in}$
$231.20\ kg/hr = \dot{m}_{in}$

$$\boxed{\dot{m}_{in} = 231.20\ kg/hr \\ \dot{m}_{out} = 431.20\ kg/hr}$$

Problem 8.25

Steam at 150 °C is fed at 75 kg/hr to a heat exchanger, where it is cooled and leaves as a mixture of steam and liquid water at 100 °C. If the cooling water comes in at 25 °C and leaves at 60 °C with a flow rate of 200 kg/hr, what is the mass fraction of steam in the outlet stream?

We begin by drawing a diagram of this process:

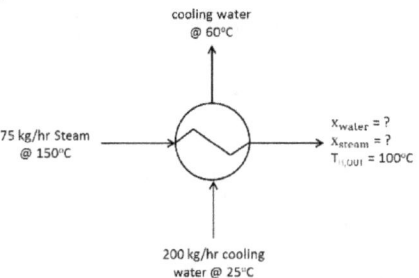

The heat gained by the cold water stream is equal to the loss in temperature of the hot stream, plus the heat of vaporization of any steam that condenses:

$$
\begin{aligned}
Q_{Lost} &= Q_{Gained} \\
\dot{m}_{hot}\Delta T_{steam}C_{steam} + \dot{m}_{hot}X_{water}L_v &= \dot{m}_{cold}\Delta T_{cold}C_{water} \\
(75 \text{ kg/hr})(150°C - 100°C)(4.184 \text{ kJ/kg °C}) + (75 \text{ kg/hr})(2260 \text{ kJ/kg}) x_{water} &= (200 \text{ kg/hr})(60°C - 25°C)(4.184 \text{ kJ/kg °C}) \\
(75 \text{ kg/hr})(50°C)(4.184 \text{ kJ/kg °C}) + (75 \text{ kg/hr})(2260 \text{ kJ/kg}) x_{water} &= (200 \text{ kg/hr})(35°C)(4.184 \text{ kJ/kg °C}) \\
15{,}690 \text{ kJ/hr} + 169{,}500 \text{ kJ/hr } (x_{water}) &= 29{,}288 \text{ kJ/hr} \\
169{,}500 \text{ kJ/hr } (x_{water}) &= 13{,}598 \text{ kJ/hr} \\
x_{water} &= 0.080
\end{aligned}
$$

We have solved for the concentration of water in the outlet stream, so:

$X_{steam} = 1 - X_{water}$

$X_{steam} = 1 - 0.080$

$X_{steam} = 0.920$

$$\boxed{X_{steam} = 0.920}$$

Chapter 8 | Material & Energy Balances

Problem 8.26

The decomposition of hydrogen peroxide into oxygen and water is exothermic, releasing 196.1 kJ/mol of H_2O_2 decomposed. 1000 mol/hr of 20°C H_2O_2 is fed at steady-state to a reactor where 20% of it decomposes.
  a. Determine the composition of the outlet stream.
  b. Determine the temperature of the outlet stream assuming no heat is lost to the surroundings. *Hint: Use appendix C.5.*

a) Determine the outlet stream composition

We start with the peroxide mole balance:

$$H_2O_2 \text{ in} - H_2O_2 \text{ consumed} = H_2O_2 \text{ out}$$
$$1000 \text{ mol/hr} - 0.2 (1000 \text{ mol/hr}) = H_2O_2 \text{ out}$$
$$800 \text{ mol/hr } H_2O_2 = H_2O_2 \text{ out}$$

H_2O mole balance:

$$H_2O \text{ generated} = H_2O \text{ out}$$
$$0.20 (1000 \text{ mol/hr } H_2O_2) \left(\frac{1 \text{ mol } H_2O}{1 \text{ mol } H_2O_2}\right) = H_2O \text{ out}$$
$$200 \text{ mol/hr } H_2O = H_2O \text{ out}$$

O_2 mole balance:

$$O_2 \text{ generated} = O_2 \text{ out}$$
$$0.20 (1000 \text{ mol/hr } H_2O_2) \left(\frac{1 \text{ mol } H_2O}{2 \text{ mol } H_2O_2}\right) = O_2 \text{ out}$$
$$100 \text{ mol/hr } O_2 = O_2 \text{ out}$$

Composition of the outlet stream:

$$y_{H2O2} = \frac{800 \text{ mol } H_2O_2}{1100 \text{ mol}}$$
$$y_{H2O2} = 0.727$$

$$y_{H2O2} = \frac{200 \text{ mol } H_2O}{1100 \text{ mol}}$$
$$y_{H2O} = 0.182$$

$$y_{O2} = \frac{100 \text{ mol } O_2}{1100 \text{ mol}}$$
$$y_{O2} = 0.091$$

$$\boxed{y_{H2O2} = 0.727; \; y_{H2O} = 0.182; \; y_{O2} = 0.091}$$

b) Determine the outlet stream temperature:

We have everything in terms of moles, so we will want to use molar heat capacities from table C.5.1. C_p of H_2O_2 is 89.091 J/mol·K, H_2O is 75.375 J/mol·K, O_2 is 29.398 J/mol·K.

$$\text{Energy generated} = \Delta \text{Energy out}$$
$$200 \text{ mol/hr } H_2O_2 (196.1 \text{ kJ/mol})(1000 \text{ J/kJ}) = \Delta T (n_{H2O2} C_{H2O2} + n_{H2O2} C_{H2O2} + n_{H2O2} C_{H2O2})$$
$$(T_F - 20°C)(800 \text{ mol/hr} \cdot 89.091 \text{ J/mol} \cdot °C +$$
$$39{,}220{,}000 \text{ J/hr} = \qquad 200 \text{ mol/hr} \cdot 75.375 \text{ J/mol} \cdot °C +$$
$$100 \text{ mol/hr} \cdot 29.398 \text{ J/mol} \cdot °C)$$
$$39{,}220{,}000 \text{ J/hr} = (T_F - 20°C)(89{,}287.6 \text{ J/hr °C})$$

$$439.25 °C = (T_F - 20°C)$$
$$459.25 °C = T_F$$

$$\boxed{\text{Exit stream temp is 459.25 °C!!}}$$

Problem 8.27

A local pond has an average volume of 10 million gallons and has a daily flow-through of 80,000 gallons via a creek. A tanker truck containing 9,000 gallons of pollutant crashes just upstream and the pollutant enters the pond. Assuming the pollutant becomes well-mixed in the pond, how long will it take for the concentration of pollutant to drop below 1.0 ppb? *Hint: consider the pond as a CSTR.*

We will treat the pond as a CSTR, meaning it is well-mixed. This allows us to treat the concentration the same everywhere. We start by putting in our constants:

	A	B
1	Initial Pollutant	9000 gal
2	Lake Volume	10000000 gal
3	Lake Flowthrough	80000 gal/day
4	dt	1 day

We can then go about building our first rows of the table:
- On the first day (day 0), all pollutant is present so we set the day zero volume in cell B11 equal to cell B1.
- In cell C11 we calculate the concentration of pollutant by dividing the pollutant volume by the lake volume.
- In cell D11, we change the pollutant volume to ppm by dividing by 10^6.
- In cell E11, we calculate the amount of pollutant to leave the lake by multiplying the concentration by the amount that leaves (flowthrough).

	A	B	C	D	E
10	Day	Volume Pollutant (gal)	Conc. Pollutant	Pollutant conc. (in ppm)	Pollutant Out (gal)
11	0	=B1	=B11/$B$2	=C11*10^6	=C11*$B$3

On next row of the table we iterate based on what has changed:
- The new volume of pollutant in the lake is the amount from the day before (cell B1) minus the pollutant that left the lake (cell E11).
- Concentration of pollutant is calculated the same way, so we do a drag-down for C12 and D12.
- The pollutant out calculation also remains the same, so do a drag-down.

	A	B	C	D	E
10	Day	Volume Pollutant (gal)	Conc. Pollutant	Pollutant conc. (in ppm)	Pollutant Out (gal)
11	0	=B1	=B11/$B$2	=C11*10^6	=C11*$B$3
12	1	=B11-E11	=B12/$B$2	=C12*10^6	=C12*$B$3

We can now drag our table down until the concentration is <1.0 ppb. **We find this occurs around day 846:**

	A	B	C	D	E
857	846	10.07203327	1.0072E-06	1.007203327	0.080576266
858	847	9.991457009	9.99146E-07	0.999145701	0.079931656

Finally we graph the results to visually watch the pollutant concentration in the lake:

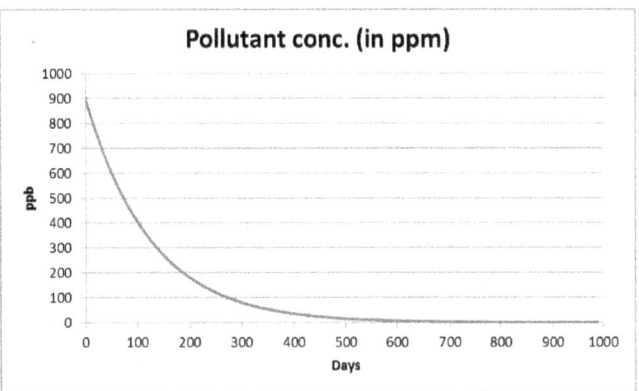

It takes approximately 846 days for the lake to drop to below 1.0 ppb

Chapter 9 Solutions
ENGINEERING STATISTICS

Problem 9.1

Given the data set below calculate the mean and standard deviation.

1.05	1.24	1.02	1.06	0.93
0.86	0.94	1.12	0.98	0.96
1	1.25	1.09	0.93	1
0.89	0.88	1.08	1.15	0.87

For larger datasets such as this, it is recommended to use a tool such as Excel.

To determine the mean, we can use the =AVERAGE() function. To determine the standard deviation, we can use the =STDEVP() function.

	A	B	C	D	E
1	1.05	1.24	1.02	1.06	0.93
2	0.86	0.94	1.12	0.98	0.96
3	1	1.25	1.09	0.93	1
4	0.89	0.88	1.08	1.15	0.87
5					
6		=AVERAGE(A1:E4)			
7		=STDEVP(A1:E4)			

	A	B	C	D	E
1	1.05	1.24	1.02	1.06	0.93
2	0.86	0.94	1.12	0.98	0.96
3	1	1.25	1.09	0.93	1
4	0.89	0.88	1.08	1.15	0.87
5					
6		1.015			
7		0.112227			

$\mu = 1.015$
$\sigma = 0.112$

Problem 9.2

Three candy bars were pulled from each of four randomly-selected boxes on the production line and weighed. Calculate standard deviation within each box, and then calculate the mean "within-box standard deviation" for the entire process.

	Box 1	Box 2	Box 3	Box 4
	2.07	2.09	2.11	2.05
	2.11	2.06	2.12	2.07
	2.06	2.07	2.06	2.08

We will solve this problem using Excel, but the general approach can be done by hand easily.

For each box, we calculate the standard deviation using =STDEVP(). We then find the mean of each of those four values using =AVERAGE().

	A	B	C	D	E	F	G
1		Box 1	Box 2	Box 3	Box 4		
2		2.07	2.09	2.11	2.05		
3		2.11	2.06	2.12	2.07		
4		2.06	2.07	2.06	2.08		mean
5	in box stdev	0.021602	0.012472	0.026247	0.012472		0.018198

> Box1 σ = 0.0216
> Box2 σ = 0.0125
> Box3 σ = 0.0.0262
> Box4 σ = 0.0125
> Mean standard deviation = 0.0182

Problem 9.3

68% of the widgets measured have a mass between 5.3 g and 8.7 g. Assuming the manufacturing process to be normally distributed, calculate the mean and standard deviation of the population.

68% of the data within a normal distribution falls within 1 sigma of the mean (total of 2 sigma):
σ = (8.7 g − 5.3 g)/2
σ = 3.4 g / 2
σ = 1.7 g

We know that the mean is 1 sigma above the lower bound (or 1 sigma below the upper bound):
μ = 5.3 g + 1.7 g
μ = 7.0 g

> μ = 7.0 g
> σ = 1.7 g

Problem 9.4

99.7% of the nails produced by a factory have a mass between 2.6 cm and 2.9 cm. Assuming the manufacturing process to be normally distributed, calculate the mean and standard deviation of the population.

99.7% of the data within a normal distribution falls within 3 sigma of the mean (total of 6 sigma):

σ = (2.9 cm − 2.6 cm)/6
σ = 0.3 cm / 2
σ = 0.15 cm

We know that the mean is 1 sigma above the lower bound (or 1 sigma below the upper bound):

μ = 2.6 cm + 0.15 cm
μ = 2.75 cm

> **μ = 2.75 cm**
> **σ = 0.15 cm**

Problem 9.5

Suppose we are sampling from an in-control process that has a normal distribution of $\mu = 400$ and $\sigma = 25$. What is the probability of collecting a sample with:

    a. A value larger than 400
    b. A value less than 325
    c. A value larger than 450

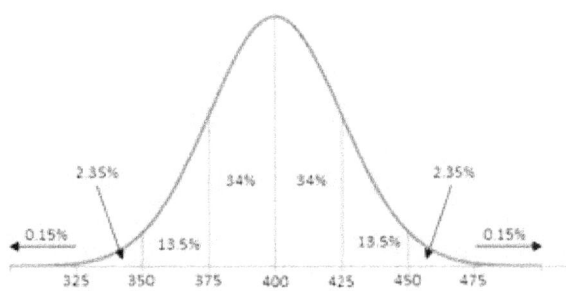

a) Value larger than 400

In a normal distribution, 50% of the data falls above the mean. Since the mean is 400, the probability of collecting a sample > 400 is 0.50.

> 400 = 0.5

b) Value less than 325

325 is 3σ less than the mean. In a normal distribution, only 0.3% of the data falls outside 6σ. Less than 325 is thus half of that value.

< 325 = 0.015

c) Value larger than 450

450 is 2σ above the mean. In a normal distribution, only 5% of the data falls outside 4σ. Greater than 450 is thus half of that value.

> 450 = 0.025

Problem 9.6

Suppose we are sampling from an in-control process that has a normal distribution of $\mu = 200$ and $\sigma = 15$. What is the probability of collecting a sample with:
- a. A value between 185 and 215
- b. A value less than 185
- c. A value greater than 170

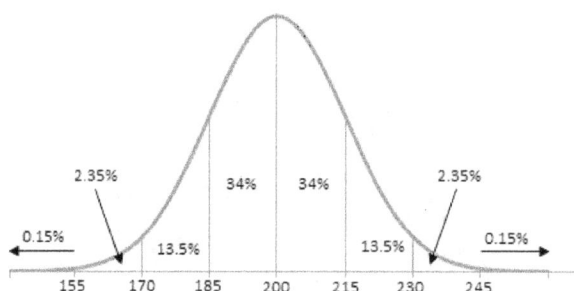

a) Value between 185 and 215

In a normal distribution, 68% of the data falls within one standard deviation of the mean.

$$185 < x < 215 = 0.68$$

b) Value less than 185

= 13.5% + 2.35% + 0.15%
= 16%

$$< 185 = 0.16$$

c) Value larger than 170

= 13.5% + 34% + 34% + 13.5% + 2.35% + 0.15%
= 97.5%

$$> 170 = 0.975$$

Problem 9.7
Is the process shown below stable? Why or why not?

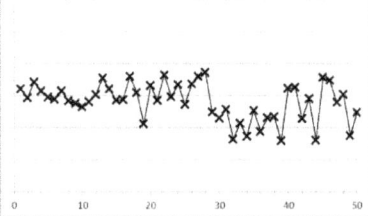

This process is **NOT STABLE**. It appears there is a trend shift around point 30, and it also may have an increase in variability around point 40.

Problem 9.8
Is the process shown below stable? Why or why not?

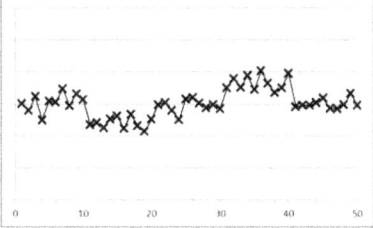

This process is **NOT STABLE**. While the variability seems constant for the entire chart, there are several distinct shifts in the process mean.

Problem 9.9
Your colleague has proposed a process change. Her data is shown below as compared to the baseline process. She argues that because her process is cheaper that the distribution means are matched, that her process should be implemented. What should you tell her?

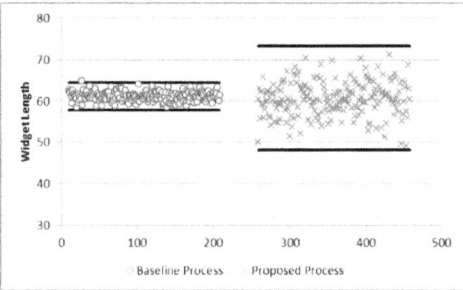

Your colleague's new process has significantly more variability than does the baseline process. We should not implement the new process because it may cause us to make parts that are outside of the desired range.

Problem 9.10

You have installed a new style of gas flow valve on one of your machines. You measured and compared the variability of the new process to the baseline process, as shown in the plot below. Should you implement this new valve on your other machines?

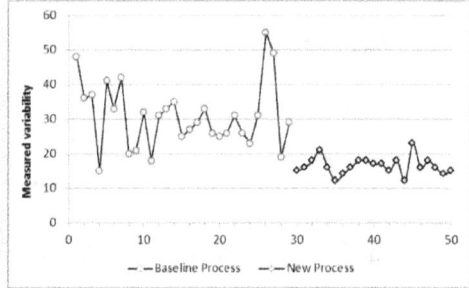

The new flow valve has significantly reduced the variability in the process. We should implement this new valve on the other machines, as it will improve the process.

Problem 9.11

Which of the chemical reactors below is performing the best?

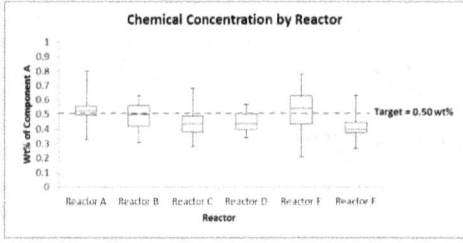

- Reactor A is close to target but has a wide variability
- Reactor B is on target and has reasonable variability
- Reactor C is off target and has higher variability than many of the other reactors
- Reactor D has a very tight distribution but is low to target
- Reactor E has a high variability
- Reactor F is off target

The best choice appears to be Reactor B

Problem 9.12

Which factory below needs to retarget its process? Which factory should attempt to better control its variability?

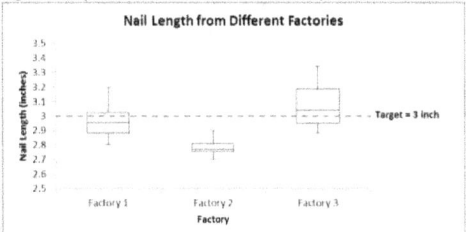

Factory 2 is producing parts far below target, and should retarget their process
Factory 3 has high variability, and should figure out where their source of variability is and fix it.

Problem 9.13

A part manufacturing process is in control and stable, with $\mu = 10$ g and $\sigma = 2$ g. Determine if the following cases are considered rare events:

    a. A part that is more than 16 g
    b. Two consecutive parts between 6 g and 8 g
    c. Two consecutive parts above 14 g
    d. Five consecutive parts less than 10 g

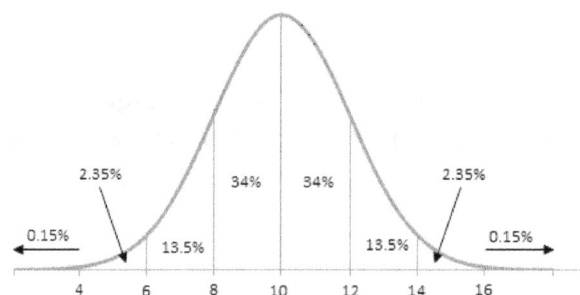

a) A part that is more than 16 g
This part would be above 3σ, which is probably of 0.0015. This is less than 0.01 so it is a rare event.

Rare event

b) Two consecutive parts between 6 g and 8 g
Between 6 g and 8 g defines the range between -1σ and -2σ, which is a probability of 0.34. 0.34*0.34 = 0.1156, this is not less than 0.01 so it is not a rare event.

Not a rare event

c) Two consecutive parts above 14 g
Being greater than 2σ is a probability of 0.025. 0.025*0.025 = 0.000625, this is less than 0.01 so it is a rare event.

Rare event

d) Five consecutive parts less than 10 g
50% of the data falls below the mean. $0.5^5 = 0.03125$, this is not less than 0.01 so it is a not rare event.

Not a rare event

Problem 9.14

A chemical mixing process is in control and stable, with product flow rates of μ = 150 L/min and σ = 15 L/min. Determine if the following cases are considered rare events:

    a. A flow less than 135 L/min
    b. Four consecutive flows are greater than 165 L/min
    c. A flow less than 135 L/min followed by a flow greater than 165 L/min
    d. Ten flows are between 135 L/min and 165 L/min

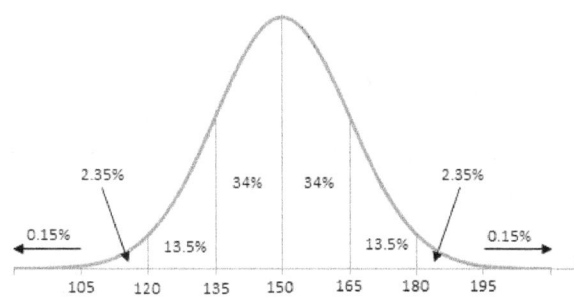

a) A flow less than 135 L/min
This flow would be below 1σ, which is probably of 0.16. This is not less than 0.01 so it is not a rare event.

    | Not a rare event |

b) Four consecutive flows are greater than 165 L/min
This flow would be above 1σ, which is probably of 0.16. $0.16^4 = 0.00066$, this is less than 0.01 so it is a rare event.

    | Rare event |

c) A flow less than 135 L/min followed by a flow greater than 165 L/min
A flow less than 135 L/min is below 1σ with a probability of 0.16, and a flow above 165 L/min is above 1σ with a probability of 0.16. $0.16 \times 0.16 = 0.032$, this is less than 0.01 and not a rare event.

    | Not a rare event |

d) Ten flows are between 135 L/min and 165 L/min
68% of the data falls within this range. $0.68^{10} = 0.0211$, this is not less than 0.01 so it is a not rare event.

    | Not a rare event |

Problem 9.15

A part thickness is measured once each day. The summary statistics were computed for the last 200 days of production, giving a μ = 19600 and σ = 160. Determine the proper control limits for the process.

LCL = μ - 3σ
LCL = 19600 − 3(160)
LCL = 19,120

UCL = μ + 3σ
UCL = 19600 + 3(160)
UCL = 20,080

Problem 9.16

A part thickness is measured once each day. The summary statistics were computed for the last 100 days of production, giving a μ = 1000 and σ = 50. Determine the proper control limits for the process.

LCL = μ - 3σ
LCL = 1000 − 3(50)
LCL = 850

UCL = μ + 3σ
UCL = 1000 + 3(50)
UCL = 1150

Problem 9.17

The overall etch rate of a sample is dependent upon two variables - the processing temperature and the speed that the sample is rotated during processing. You run an experiment where you modified each variable, and plotted the results below. Based on your data, what conclusions can you make about the impact of each process variable?

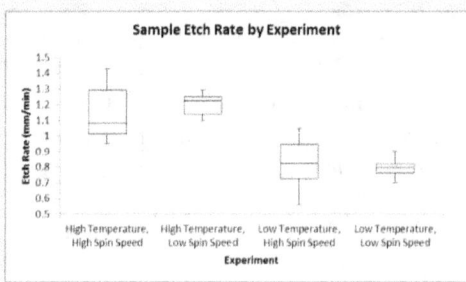

One way to evaluate each variable is to break them out in a table as per below. This allows us to isolate the effects of each variable:

	Low Speed	High Speed
Low Temp	low etch ate low variability	low etch rate high variability
High Temp	high etch rate low variability	high etch rate high variability

- It appears that by changing the speed, we change the variability. Changing the spin speed does not appear to have any effect on the etch rate.
- The temperature appears to affect the etch rate, and has no observable effect on the variability.

Chapter 10 Solutions
COMPUTER ENGINEERING

Problem 10.1
Evaluate the expression X = A + B • C for A = 1, B = 0, and C = 1.

X = A + B • C
X = 1 + 0 • 1
X = 1 + 0
X = 1 (True)

| True |

Problem 10.2
Evaluate the expression X = A + B • C' for A = 1, B = 1, and C = 1.

X = A + B • C'
X = 1 + 1 • 1'
X = 1 + 1 • 0
X = 1 + 0
X = 1 (True)

| True |

Problem 10.3
Evaluate the expression X = A • B' • (A + C) for A = 1, B = 0, and C = 0.

X = A • B' • (A + C)
X = 1 • 0' • (1 + 0)
X = 1 • 1 • 1
X = 1 (True)

| True |

Problem 10.4
Evaluate the expression X = A' • (B + C) • D for A = 1, B = 0, C = 1, and D = 1.

X = A' • (B + C) • D
X = 1' • (0 + 1) • 1
X = 0 • 1 • 1
X = 0 (False)

| False |

Problem 10.5
Consider the circuit below and the end variable L = "the light is on", with input variables A = "the switch is closed" and B = "the switch is closed". Express the relationship as a logic equation for L in terms of A and B.

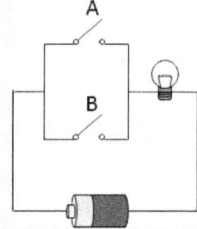

The light will turn on if either of the parallel switches (A or B) is closed, regardless of the state of the other switch. Thus this is an OR statement.

Answer: L = A + B

Problem 10.6
Consider the circuit below and the end variable L = "the light is on", with input variables A,B, and C = "the switch is closed". Express the relationship as a logic equation for L in terms of A, B, and C.

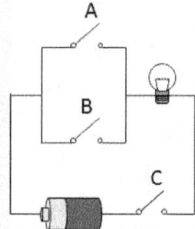

The light will turn on only if the C is closed, and only if at least one of the parallel switches (A or B) is closed. Thus this is a two operation statement using AND and OR.

Answer: L = (A + B) • C

Problem 10.7
Build a truth table for the expression X = A • B'.

Input		Output
A	B	A • B'
1	1	0
1	0	1
0	1	0
0	0	0

Problem 10.8
Build a truth table for the expression X = A + B'.

Input		Output
A	B	A + B'
1	1	1
1	0	1
0	1	0
0	0	1

Problem 10.9
Build a truth table for the expression X = (A + B') • C.

Input			Output
A	B	C	(A + B') • C
1	1	1	1
1	1	0	0
1	0	1	1
1	0	0	0
0	1	1	0
0	1	0	0
0	0	1	1
0	0	0	0

Problem 10.10
Build a truth table for the expression X = (A + B) • (A + C).

Input			Output
A	B	C	(A + B) • (A + C)
1	1	1	1
1	1	0	1
1	0	1	1
1	0	0	1
0	1	1	1
0	1	0	0
0	0	1	0
0	0	0	0

Problem 10.11
Convert the following binary numbers to their equivalent in decimal:
    a. 1001
    b. 10111
    c. 11000101

a) $= (1 \cdot 2^3) + (0 \cdot 2^2) + (0 \cdot 2^1) + (1 \cdot 2^0)$
    $= 8 + 0 + 0 + 1$
    = 9

b) $= (1 \cdot 2^4) + (0 \cdot 2^3) + (1 \cdot 2^2) + (1 \cdot 2^1) + (1 \cdot 2^0)$
    $= 16 + 0 + 4 + 2 + 1$
    = 23

c) $= (1 \cdot 2^7) + (1 \cdot 2^6) + (0 \cdot 2^5) + (0 \cdot 2^4) + (0 \cdot 2^3) + (1 \cdot 2^2) + (0 \cdot 2^1) + (1 \cdot 2^0)$
    $= 128 + 64 + 0 + 0 + 0 + 4 + 0 + 1$
    = 197

Problem 10.12
Convert the following binary numbers to their equivalent in decimal:
- a. 1000011
- b. 1100011
- c. 11111100

a) $= (1 \cdot 2^6) + (0 \cdot 2^5) + (0 \cdot 2^4) + (0 \cdot 2^3) + (0 \cdot 2^2) + (1 \cdot 2^1) + (1 \cdot 2^0)$
$= 64 + 0 + 0 + 0 + 0 + 2 + 1$
= 67

b) $= (1 \cdot 2^6) + (1 \cdot 2^5) + (0 \cdot 2^4) + (0 \cdot 2^3) + (0 \cdot 2^2) + (1 \cdot 2^1) + (1 \cdot 2^0)$
$= 64 + 32 + 0 + 0 + 0 + 2 + 1$
= 99

c) $= (1 \cdot 2^7) + (1 \cdot 2^6) + (1 \cdot 2^5) + (1 \cdot 2^4) + (1 \cdot 2^3) + (1 \cdot 2^2) + (0 \cdot 2^1) + (0 \cdot 2^0)$
$= 128 + 64 + 32 + 16 + 8 + 4 + 0 + 0$
= 252

Problem 10.13
Convert the following decimal numbers to their equivalent in binary:
- a. 10
- b. 64
- c. 127

a) 10 = **1010**
b) 64 = **1000000**
c) 127 = **11111110**

Problem 10.14
Convert the following decimal numbers to their equivalent in binary:
- a. 13
- b. 20
- c. 254

a) 13 = **1101**
b) 20 = **10100**
c) 254 = **11111110**

Problem 10.15
What number (in decimal) is represented by the circuit below?

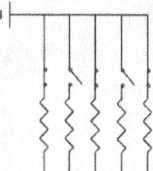

From left to right, the values are 0, 1, 0, 1, 0. This correlates to:
$$= (1 \cdot 2^4) + (0 \cdot 2^3) + (1 \cdot 2^2) + (0 \cdot 2^1) + (1 \cdot 2^0)$$
$$= 16 + 0 + 4 + 0 + 1$$
$$= \mathbf{21}$$

Problem 10.16
What number (in decimal) is represented by the circuit below?

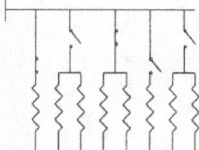

From left to right, the values are 1, 0, 0, 1, 1, 1, 0, 0. This correlates to:
$$= (1 \cdot 2^7) + (0 \cdot 2^6) + (0 \cdot 2^5) + (1 \cdot 2^4) + (1 \cdot 2^3) + (1 \cdot 2^2) + (0 \cdot 2^1) + (0 \cdot 2^0)$$
$$= 128 + 0 + 0 + 16 + 8 + 4 + 0 + 0$$
$$= \mathbf{156}$$

Problem 10.17
A specific password has 16-bit encryption. How many possible password combinations are there?

There will be 2^n combinations, or 2^{16} combinations. 2^{16} = 65,536 combinations.

65,536 combinations

Problem 10.18
A hard drive has 128-bit encryption. If a hacker's supercomputer can enter 10 codes per second, how long would it take the hacker to try all combinations? Assume there are no resets or delays in between the individual attempts.

There will be 2^n combinations, or 2^{128} combinations. 2^{128} = 3.4028 x 10^{38} combinations.

3.4028 x 10^{38} ~~codes~~	~~sec~~	~~min~~	~~hr~~	~~day~~	year	=	= 1.078 x 10^{30} years
	10 ~~code~~	60 ~~sec~~	60 ~~min~~	24 ~~hr~~	365.25 day		

Chapter 11 Solutions
RELIABILITY ENGINEERING

Problem 11.1
100 devices were tested for a period of 500 hours. After 200 hours 2 devices had failed, and at the end of testing a total of 7 devices had failed. Determine R(200) and R(500).

To find the reliability over each time, we apply the reliability formula:

$$R(200) = \frac{n_s(t)}{n} \qquad\qquad R(500) = \frac{n_s(t)}{n}$$

$$R(200) = \frac{98}{100} \qquad\qquad R(500) = \frac{93}{100}$$

R(200) = 0.98 $\qquad\qquad$ **R(500) = 0.93**

$$\boxed{R(200) = 0.98 \;;\; R(500) = 0.93}$$

Problem 11.2
30 devices were tested for a period of 20 hours. At the end of testing 6 devices had failed. What is the probability of device failure for the testing period?

The failure probability over time t is given by:

$$F(20) = \frac{n_f(t)}{n}$$

$$F(20) = \frac{6}{30}$$

F(20) = 0.2

$$\boxed{F(20) = 0.2}$$

Problem 11.3

A company produces 10,000 computer chips per day. 500 are tested for a period of 1000 hours, and during testing a total of 7 failed: two at 50 hours, two at 100 hours, two at 200 hours, and one at 700 hours. Determine the failure rate and MTBF for a chip.

First calculate the total operating time:
$T_O = T_T -$ downtime
$T_O = 500(1000) - [\,2(950) + 2(900) + 2(800) + 300\,]$
$T_O = 494{,}400$ hrs

Find the failure rate:
$$\lambda = \frac{\text{failures}}{\text{total operating time}} = \frac{7}{494{,}400 \text{ hrs}}$$
$\lambda = 1.41 \times 10^{-5}$ fails/hr

Find the MTBF:
$$\text{MTBF} = \frac{\text{total operating time (hrs)}}{\text{number of failures}} = \frac{494{,}400 \text{ hrs}}{7}$$
MTBF = 70,628.57 hr/fail

> $\lambda = 1.41 \times 10^{-5}$ fails/hr ; MTBF = 70,628.57 hr/fail

Problem 11.4

300 of the chips from problem 3 of this chapter are used to build a supercomputer. If failure of any one chip brings down the entire system, how many system failures can be expected per month? What is the MTBF for the supercomputer?

Convert the fail rate into fails/month:
$$\frac{1.41 \times 10^{-5} \text{ fails}}{\text{hr}} \cdot \frac{24 \text{ hr}}{\text{day}} \cdot \frac{30 \text{ day}}{\text{month}}$$
$\lambda_{chip} = 1.1019 \times 10^{-2}$ fail/month

Since the chips are all in series:
$\lambda_{computer} = 1.1019 \times 10^{-2}$ fail/month/unit $\cdot$ 300 units
$\lambda_{computer} = 3.05$ fail/month

Determine the MTBF:
$$\text{MTBF} = \frac{1}{\lambda} = \frac{1}{3.05}$$
MTBF = 0.327 months = ~9.8 days

> $\lambda = 3.05$ fails/mo ; MTBF = 9.8 days

Problem 11.5
A car has a failure rate of 10^{-4} per mile traveled. What is the probability of successfully traveling 1000 miles?

$R(t) = e^{-(\lambda t)}$
$R(1000) = e^{-(0.0001 \cdot 1000)}$
$R(1000) = e^{-(0.01)}$
$R(1000) = 0.905$

$\boxed{R(1000) = 0.905}$

Problem 11.6
A two-component system is completely non-redundant as shown below. Over a 1-year period, component A failed 15 times and component B failed 3 times. Component A takes 4 hours to replace, and component B takes 18 hours to replace. What is the MTBF for component A, for component B, and for the overall system?

We first calculate the total operational time:
$T_O = 365 \cdot 24 - 15 \cdot 4 - 3 \cdot 18$
$T_O = 8646$ hrs

We use that to calculate the MTBF for each component and the system:

$$MTBF_A = \frac{T_O}{\text{\# of failures of A}}$$
$$MTBF_A = \frac{8646}{15}$$
$$\boxed{MTBF_A = 576.4 \text{ hrs}}$$

$$MTBF_B = \frac{T_O}{\text{\# of failures of B}}$$
$$MTBF_B = \frac{8646}{3}$$
$$\boxed{MTBF_B = 2882 \text{ hrs}}$$

$$MTBF_S = \frac{T_O}{\text{\# of system failures}}$$
$$MTBF_S = \frac{8646}{18}$$
$$\boxed{MTBF_S = 480.33 \text{ hrs}}$$

Problem 11.7

Two identical components are connected in full redundancy as shown below. Over a 1-year period, there were 3 part failures, none simultaneous. Repair time for each failure is 5 days. What is the MTBF of the part and of the two-component system?

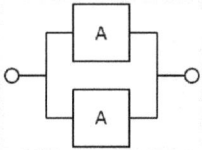

We first calculate the total operational time:
$T_O = 2 \cdot 365 \cdot 24 - 3 \cdot 5 \cdot 24$
$T_O = 25{,}920$ hrs

We use that to calculate the MTBF for each component and the system:

$$\text{MTBF}_A = \frac{T_O}{\text{\# of failures of A}} \qquad \text{MTBF}_S = \frac{T_O}{\text{\# of system failures}}$$

$$\text{MTBF}_A = \frac{25{,}920}{3} \qquad \text{MTBF}_S = \frac{8646}{0}$$

$$\boxed{\text{MTBF}_A = 8640 \text{ hrs}}$$

The system does not fail within the time period, there is no measurable system MTBF.

Problem 11.8

In problem 6 of this chapter, what is the 1-month reliability of each component and of the overall system?

We use the MTBF as found in problem 6 to calculate the time-based reliabilities. We first must convert the MTBF from days to months:
MTBF$_A$ = 576.4 hrs → 0.8005 months
MTBF$_B$ = 576.4 hrs → 4.0027 months
MTBF$_S$ = 576.4 hrs → 0.6771 months

$$R(t) = e^{-(t/\text{MTBF})} \qquad R(t) = e^{-(t/\text{MTBF})} \qquad R(t) = e^{-(t/\text{MTBF})}$$

$$R(t) = e^{-(1\text{ month}/0.8005\text{ month})} \qquad R(t) = e^{-(1\text{ month}/4.0027\text{ month})} \qquad R(t) = e^{-(1\text{ month}/0.6771\text{ month})}$$

$$\boxed{R_A(1 \text{ mo}) = 0.287} \qquad \boxed{R_B(1 \text{ mo}) = 0.779} \qquad \boxed{R_S(1 \text{ mo}) = 0.228}$$

Problem 11.9
In problem 7 of this chapter, what is the 1-month reliability of the component and of the system?

We use the MTBF as found in problem 6 to calculate the time-based reliabilities. We first must convert the MTBF from days to months:
$MTBF_A$ = 8460 hrs → 0.8005 months
$MTBF_S$ = undeterminable

$$R(t) = e^{-(t/MTBF)}$$
$$R(t) = e^{-(1\,month/11.75\,month)}$$
$$\boxed{R_A(1\,mo) = 0.918}$$

Because we could not determine the MTBF of the system, we must use the 1-month reliability to determine the system reliability. The components are in parallel so:

$R_S = 1 - (1 - R_A)^2$
$R_S = 1 - (1 - 0.918)^2$
$R_S = 1 - (0.082)^2$
$R_S = 1 - 0.006724$
$R_S = 0.993$

$\boxed{R_S = 0.993}$

Problem 11.10
Determine the reliability of the below system.

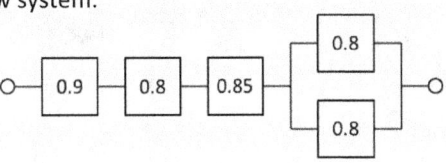

The set of redundant components have a combined reliability of:
$R_A = [1 - (1 - 0.8)^2]$
$R_A = [1 - (0.2)^2]$
$R_A = 0.96$

All components are then in series:
$R_S = (0.9)(0.8)(0.85)(0.96)$
$R_S = 0.587$

$\boxed{R_S = 0.587}$

Problem 11.11
Determine the reliability of the below system.

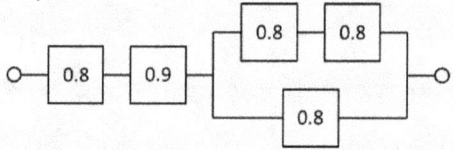

The first branch has a reliability of:
$R_A = (0.8)^2$
$R_A = 0.64$

So the parallel components have a reliability of:
$R_B = 1 - (1 - 0.8)(1 - 0.64)$
$R_B = 1 - (0.2)(0.36)$
$R_B = 0.928$

We then have two components in series with each of the two parallel branches:
$R_S = (0.8)(0.9)(0.928)$
$R_S = 0.668$

$\boxed{R_S = 0.668}$

Problem 11.12
Determine the reliability of the below system.

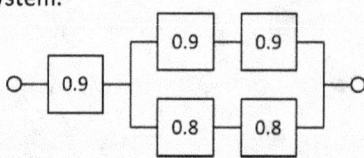

The first branch has a reliability of:
$R_A = (0.9)(0.9)$
$R_A = 0.81$

The second branch has a reliability of:
$R_B = (0.8)(0.8)$
$R_B = 0.64$

We then have one component in series with each of the two parallel branches:
$R_S = (0.9)[1 - (1 - 0.81)(1 - 0.64)]$
$R_S = (0.9)[1 - (0.19)(0.36)]$
$R_S = (0.9)(0.9316)$
$R_S = 0.838$

$\boxed{R_S = 0.838}$

Problem 11.13
Determine the reliability of the below system.

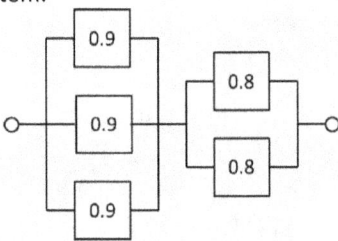

We have two groups of parallel components that are connected in series:
$R_S = [1 - (1 - 0.9)^3] [1 - (1 - 0.8)^2]$
$R_S = [1 - (0.1)^3] [1 - (0.2)^2]$
$R_S = [1 - 0.001] [1 - 0.04]$
$R_S = [0.999] [0.96]$
$R_S = 0.959$

$\boxed{R_S = 0.959}$

Problem 11.14
A system design requires that 100 identical components be connected in series. The overall system reliability must be at least 0.99. What is the minimum allowed reliability for each component?

$$R_S = R_c^N$$
$$0.99 = R_c^{100}$$
$$0.99^{(1/100)} = R_c$$
$$0.9999 = R_c$$

$\boxed{\text{Minimum reliability of each component is } 0.9999}$

Problem 11.15
A system design requires that 10 identical components be connected in parallel. The overall system reliability must be at least 0.99. What is the minimum allowed reliability for each component?

$$R_S = 1 - (1 - R_c)^N$$
$$0.99 = 1 - (1 - R_c)^{10}$$
$$0.01 = (1 - R_c)^{10}$$
$$(0.01)^{(1/10)} = 1 - R_c$$
$$0.6310 = 1 - R_c$$
$$0.3690 = R_c$$

$\boxed{\text{Minimum reliability of each component is } 0.3690}$

Problem 11.16

A series system has identical components each with reliability of 0.998. What is the maximum number of components that are allowed in order for the system reliability to be no less than 0.90?

$$R_S = R_c^N$$
$$0.90 = (0.998)^N$$
$$\log(0.90) = N \log(0.998)$$
$$\log(0.90) / \log(0.998) = N$$
$$52.627 = N$$

No more than 52 components are allowed

Problem 11.17

A parallel system has identical components each with reliability of 0.6. What is the minimum number of components required for the system reliability to be at least 0.90?

$$R_S = 1 - (1 - R_c)^N$$
$$0.90 = 1 - (1 - 0.6)^N$$
$$0.10 = (0.4)^N$$
$$\log(0.10) = N \log(0.4)$$
$$\log(0.10) / \log(0.4) = N$$
$$2.51 = N$$

Need at least 3 components

Problem 11.18
Derive an expression for the below six-component system in terms of the original components.

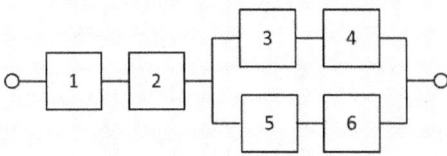

First combine 3 and 4 into equivalent component 7:
$R_7 = R_3 R_4$

Next combine 5 and 6 into equivalent component 8:
$R_8 = R_5 R_6$

Combine 7 and 8 into equivalent component 9:
$R_9 = 1 - (1 - R_7)(1 - R_8)$
$R_9 = 1 - (1 - R_7 - R_8 + R_7 R_8)$
$R_9 = R_7 + R_8 - R_7 R_8$

The remaining components are in series:
$R_S = R_1 R_2 R_9$
$R_S = R_1 R_2 R_7 + R_1 R_2 R_8 - R_1 R_2 R_7 R_8$
$R_S = R_1 R_2 R_3 R_4 + R_1 R_2 R_5 R_6 - R_1 R_2 R_3 R_4 R_5 R_6$

$$\boxed{R_S = R_1 R_2 R_3 R_4 + R_1 R_2 R_5 R_6 - R_1 R_2 R_3 R_4 R_5 R_6}$$

Problem 11.19
Derive an expression for the below five-component system in terms of the original components.

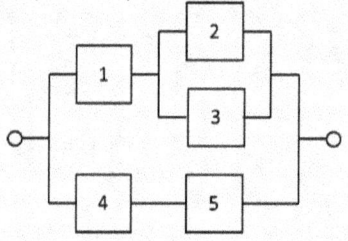

Combine 2 and 3 into equivalent component 6:
$R_6 = 1 - (1 - R_2)(1 - R_3)$
$R_6 = R_2 + R_3 - R_2R_3$

Combine 1 and 6 into equivalent component 7:
$R_7 = R_1R_6$
$R_7 = R_1R_2 + R_1R_3 - R_1R_2R_3$

Combine 4 and 5 into equivalent component 8:
$R_8 = R_4R_5$

Component 7 and component 8 are in parallel:
$R_S = 1 - (1 - R_7)(1 - R_8)$
$R_S = R_7 + R_8 - R_7R_8$
$R_S = R_1R_2 + R_1R_3 - R_1R_2R_3 + R_4R_5 - R_1R_2R_4R_5 - R_1R_3R_4R_5 + R_1R_2R_3R_4R_5$

$$\boxed{R_S = R_1R_2 + R_1R_3 - R_1R_2R_3 + R_4R_5 - R_1R_2R_4R_5 - R_1R_3R_4R_5 + R_1R_2R_3R_4R_5}$$

Problem 11.20
Derive an expression for the below four-component system in terms of the original components.

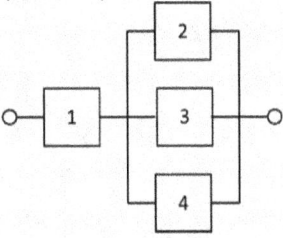

Combine 2 and 3 and 4 into equivalent component 5:
$R_5 = 1 - (1 - R_2)(1 - R_3)(1 - R_4)$
$R_5 = 1 - (1 - R_2 - R_3 - R_2R_3)(1 - R_4)$
$R_5 = 1 - (1 - R_2 - R_3 - R_2R_3 - R_4 + R_2R_4 + R_3R_4 - R_2R_3R_4)$
$R_5 = R_2 - R_3 - R_2R_3 + R_4 - R_2R_4 - R_3R_4 + R_2R_3R_4$

Components 1 and 5 are in series:
$R_S = R_1R_5$
$R_S = R_1R_2 - R_1R_3 - R_1R_2R_3 + R_1R_4 - R_1R_2R_4 - R_1R_3R_4 + R_1R_2R_3R_4$

$$\boxed{R_S = R_1R_2 - R_1R_3 - R_1R_2R_3 + R_1R_4 - R_1R_2R_4 - R_1R_3R_4 + R_1R_2R_3R_4}$$

Problem 11.21
A system consists of three components in series, each with a reliability of 0.95. A second set of components is purchased. Determine the system reliability if the components are connected in:
  a. Low-level redundancy
  b. High-level redundancy

In low-level redundancy, there are two redundant 3-component chains, so:
$R_{LL} = 1 - (1 - (0.95)^3)^2$
$R_{LL} = 1 - (1 - 0.857)^2$
$R_{LL} = 1 - (0.143)^2$
$R_{LL} = 0.980$

In high-level redundancy, there are three redundant components connected in series, so:
$R_{HL} = [1 - (1 - 0.95)^2]^3$
$R_{HL} = [1 - (0.05)^2]^3$
$R_{HL} = [1 - 0.0025]^3$
$R_{HL} = 0.993$

$$\boxed{\text{low-level} = 0.980; \text{ high-level} = 0.993}$$

Problem 11.22

A device consists of 100 switches. In branch A, 20 are connected in series. Branch A is connected in series to a set of parallel branches B and C. Branch B contains 20 switches connected in series, and Branch C contains 60 switches connected in series. All switches are identical and have a reliability of 0.98. Determine the system reliability.

Reliability of branch A:
$R_A = (0.98)^{20}$
$R_A = 0.668$

Reliability of branch B:
$R_B = (0.98)^{20}$
$R_B = 0.668$

Reliability of branch C:
$R_C = (0.98)^{60}$
$R_C = 0.298$

System reliability:
$R_S = R_A \cdot (1 - (1 - R_B)(1 - R_C))$
$R_S = 0.668 \cdot (1 - (1 - 0.668)(1 - 0.298))$
$R_S = 0.668 \cdot (1 - (0.332)(0.702))$
$R_S = 0.668 \cdot (0.767)$
$R_S = 0.512$

$\boxed{R_S = 0.512}$

Problem 11.23

Unplanned disruptions in the supply of power to your factory are extremely disruptive and cause loss of income. The power supply system contains two components connected in parallel. You are the system owner and tasked with ensuring constant service to the factory through pre-emptive replacement of the components. Given the testing data below collected on 1000 components over 1000 hours, propose a change frequency that will get system reliability to 0.9999.

Time (hrs)	# failures
0-100	2
100-200	1
200-300	2
300-400	0
400-600	3
600-800	4
800-1000	15

The system must have a reliability of 0.9999, so we determine the required reliability for each component:

$$R_s = 1 - (1 - R_c)^N$$
$$0.9999 = 1 - (1 - R_c)^2$$
$$0.0001 = (1 - R_c)^2$$
$$(0.0001)^{1/2} = ((1 - R_c)^2)^{1/2}$$
$$0.01 = 1 - R_c$$
$$0.99 = R_c$$

Each component must have a reliability of 0.99. Going to our testing data, we must determine where the component reliability dips below 0.99:

Time (hrs)	# failures	R(t)
0-100	2	$\frac{998}{1000} = 0.998$
100-200	1	$\frac{997}{1000} = 0.997$
200-300	2	$\frac{995}{1000} = 0.995$
300-400	0	$\frac{995}{1000} = 0.995$
400-600	3	$\frac{992}{1000} = 0.992$
600-800	4	$\frac{988}{1000} = 0.988$

The reliability dips below 0.99 after 600 hours. Thus, we should change out our components every 600 hours.

Change each component every 600 hours

Chapter 12 Solutions
MATERIALS SCIENCE AND ENGINEERING

Problem 12.1
A cylindrical sample of a material undergoes completely elastic deformation under a shearing stress. From the diagrams below select the one that would best reflect the sample's final shape.

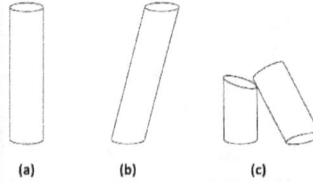

Samples (b) and (c) both show permanent (plastic) deformation. Sample A has returned to its original shape, indicating it was completely elastic deformation.

Sample A

Problem 12.2
A cylindrical sample of a material fractures under a shearing stress. From the diagrams below select the one that would best reflect the sample's final shape.

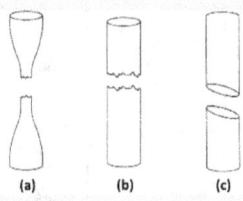

All three samples have fractured, but sample (a) and sample (b) appear to have done so under a tensile stress. Sample (c) is the only one that appears to have been under a shearing stress.

Sample C

Problem 12.3

A cylindrical sample of a material with low toughness is dropped and impacts the floor. From the diagrams below select the one that would best reflect the sample's final shape.

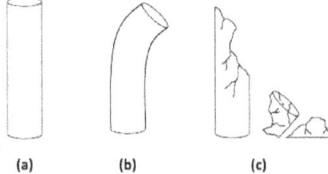

(a)   (b)   (c)

Low toughness indicates that the sample is likely to break under an impact. Sample (a) shows no deformation, sample (b) shows some deformation, but sample (c) has completely fractured.

> **Sample C**

Problem 12.4

A metal rod is 0.5 m long and has a diameter of 10 mm. Under a force of 50 kN it lengthens by 5 mm. Determine the engineering stress and the engineering strain in the rod.

First convert to SI units:
10 mm = 0.010 m
5 mm = 0.005 m

To calculate the engineering stress:
$$\sigma = \frac{F}{A_o}$$
$$\sigma = \frac{50 \text{ kN}}{\pi(0.005\text{m})^2} = \frac{50{,}000 \text{ N}}{7.85 \times 10^{-5} \text{ m}^2}$$
$$\sigma = 636.94 \times 10^6 \text{ Pa}$$

$$\varepsilon = \frac{\Delta L}{L_o}$$
$$\varepsilon = \frac{0.005 \text{ m}}{0.5 \text{ m}}$$
$$\varepsilon = 0.1$$

> $\sigma = 636.9$ MPa
> $\varepsilon = 0.1$

Problem 12.5
A steel bar is 0.5 m long and 5 mm in diameter. It is stretched by 0.06 mm under a force of 3 kN. Determine the engineering stress and the engineering strain in the rod.

First convert to SI units:
6 mm = 0.006 m
5 mm = 0.005 m

To calculate the engineering stress:
$$\sigma = \frac{F}{A_o}$$
$$\sigma = \frac{30 \text{ kN}}{\pi(0.0025\text{m})^2} = \frac{30,000 \text{ N}}{7.85 \times 10^{-5} \text{ m}^2}$$
$$\sigma = 1527.88 \times 10^6 \text{ Pa}$$

$$\varepsilon = \frac{\Delta L}{L_o}$$
$$\varepsilon = \frac{0.006 \text{ m}}{0.5 \text{ m}}$$
$$\varepsilon = 0.12$$

$\boxed{\sigma = 1527.9 \text{ MPa} \\ \varepsilon = 0.12}$

Problem 12.6
A 240 mm long metal rod is 80 mm in diameter. If the bar is stressed to 280 MPa in the elastic region where the modulus of elasticity is 205 GPa, determine the resulting strain and the applied force. 0.00136, 1.407 MN

First convert to SI units:
240 mm = 0.24 m
80 mm = 0.08 m

The applied force can be found by:
$F = \sigma \cdot A_o$
$F = 280 \times 10^6 \text{ Pa} \cdot \pi(0.04)^2$
$F = 1.407 \times 10^6 \text{ N}$

Because this is in the elastic region, we can related stress to strain with the elastic modulus:
$\sigma = E \varepsilon$
$$\varepsilon = \frac{\sigma}{E} = \frac{280 \text{ MPa}}{206 \text{ GPa}} = \frac{280 \times 10^6}{206 \times 10^9}$$
$\varepsilon = 0.00136$

$\boxed{F = 1.407 \times 10^6 \text{ N} \\ \varepsilon = 0.00136}$

Problem 12.7

The stress-strain curve for a brittle ceramic material is shown below. What will happen to the material if it experiences a tensile stress of a) 150 MPa, b) 300 Mpa, and c) 350 Mpa?

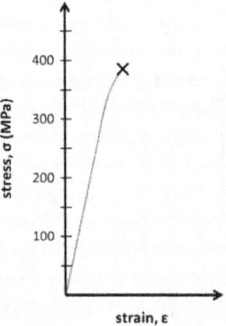

a) At 150 MPa, the stress is in the elastic region and thus the material undergoes elastic deformation.

b) At 300 MPa, the stress is still in the elastic region and thus the material undergoes elastic deformation

c) At 350 MPa, the stress is now in the plastic region and thus the material undergoes permanent deformation

Problem 12.8

The stress-strain curve for a ductile metallic material is shown below. What will happen to the material if it experiences a tensile stress of a) 25 Mpa, b) 100 MPa, and c) 125 Mpa?

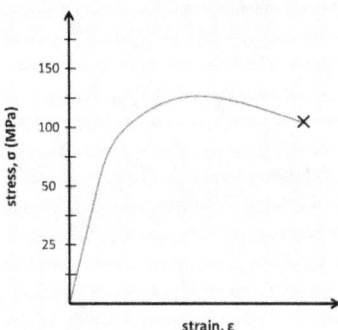

a) At 25 MPa, the stress is in the elastic region and thus the material undergoes elastic deformation.

b) At 100 MPa, the stress is now in the plastic region and thus the material undergoes permanent deformation

c) At 125 MPa, the material fractures

Problem 12.9

The stress-strain curve for a material is shown below. Estimate the resulting deformation to the material if it experiences a tensile stress of a) 75 MPa, b) 175 MPa, and c) 200 MPa?

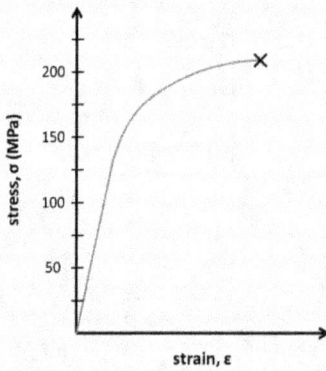

When the stress is unloaded, the unloading occurs linearly and parallel to the loading line:

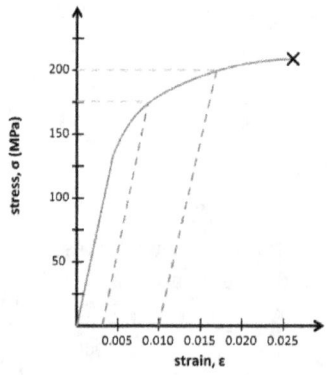

a) At 75 MPa, the stress is in the elastic region and thus there is no permanent deformation
b) At 175 MPa, the resulting deformation is ~ 0.003
c) At 200 MPa, the resulting deformation is ~0.010

Problem 12.10

The stress-strain curve for a material is shown below. Estimate the resulting deformation to the material if it experiences a tensile stress of a) 30 MPa, b) 70 MPa, and c) 80 MPa?

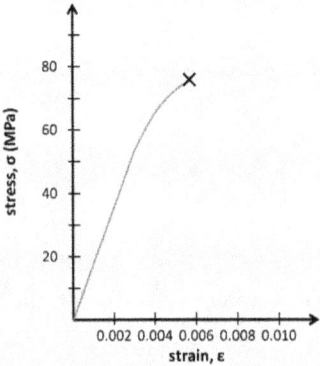

When the stress is unloaded, the unloading occurs linearly and parallel to the loading line:

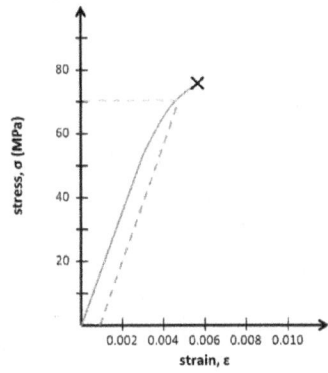

a) At 30 MPa, the stress is in the elastic region and thus there is no permanent deformation
b) At 70 MPa, the resulting deformation is ~ 0.001
c) At 80 MPa, the material fractures

Problem 12.11

Two cylindrical rods are made from the same material, but rod A has twice the radius of rod B. Which rod can withstand a greater load before failure?

Rod A has a larger cross-section of area to carry the load, due to its geometry. Rod A can thus carry more load before it breaks.

Problem 12.12

Two rectangular samples are made from the same material, but sample A is half the width of sample B. Which sample can withstand a greater stress before failure?

The maximum stress is a material property, and is the same in both rods when they break. Both samples break at the same stress.

Problem 12.13

Iodine forms a crystal structure with a unit cell known as the base centered orthorhombic, as shown below. How many atoms are in this unit cell?

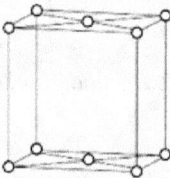

The base centered orthorhombic structure has two face atoms each contributing $1/2$ of an atom each and eight corner atoms each contributing $1/8$ of an atom each. The total is:

face atoms:	$2 \times 1/2$	=	1
corner atoms:	$8 \times 1/8$	=	1
Total:		=	2

There are two atoms in the unit cell

Problem 12.14

Bismuth forms a crystal structure with a unit cell known as the base centered monoclinic, as shown below. How many atoms are in this unit cell?

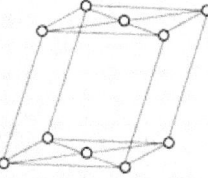

The base centered monoclinic structure has two face atoms each contributing $1/2$ of an atom each and eight corner atoms each contributing $1/8$ of an atom each. The total is:

face atoms:	$2 \times 1/2$	=	1
corner atoms:	$8 \times 1/8$	=	1
Total:		=	2

There are two atoms in the unit cell

Problem 12.15

Determine the number of anions and the number of cations in the zinc blende structure unit cell, and prove that the structure has a net charge of zero for ions of equal but opposite charge.

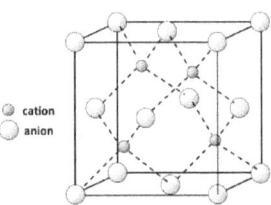

The zinc blende structure has eight corner anions each contributing $1/8$ of an anion each and six face anions contributing $1/2$ of an anion each. The total # of anions is:

corner anions:	$8 \times 1/8$	=	1
face anions	$6 \times 1/2$	=	3
Total:		=	4

The zinc blende structure has four interior cations each contributing a full cation each. The total # of cations is 4.

In this structure the anions and cations have equal and opposite charges. The net charge of the structure is thus:

# anions:	4 x -X	=	-4X
# cations:	4 x +X	=	+4X
net charge:		=	0

There are 4 anions and 4 cations, leading to a net charge of zero

Problem 12.16

Determine the number of anions and the number of cations in the perovskite structure unit cell, and prove that the structure has a net charge of zero for anions of -2 charge, a center cation of +4 charge, and corner cations of +2 charge.

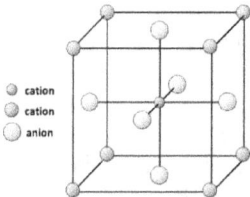

The perovskite structure has six face anions each contributing $1/2$ of an anion each. The total # of anions is 3.

The perovskite structure has one interior cation, and eight corner cations each contributing $1/8$ of an cation each.

The net charge of the structure is thus:

# anions:	3 x -2	=	-6
# center cations:	1 x +4	=	+4
# corner cations:	1 x +2	=	+2
net charge:		=	0

There are 3 anions and one of each cation, leading to a net charge of zero

Problem 12.17

Determine the number of anions and the number of cations in the fluorite structure unit cell, and prove that the structure has a net charge of zero for cations of +2 charge and anions of -1 charge.

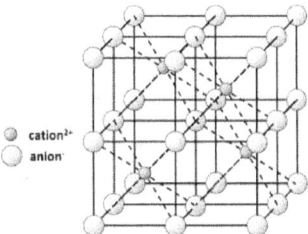

The fluorite structure has eight corner anions each contributing $1/8$ of an anion each, six face anions contributing $1/2$ of an anion each, twelve edge anions contributing $1/4$ of an anion each, and one center anion. The total # of anions is:

corner anions:	8 x $1/8$	=	1
face anions	6 x $1/2$	=	3
edge anions:	12 x $1/4$	=	3
center anions:	1 x 1	=	1
Total:		=	8

The fluorite structure has four interior cations each contributing a full cation each. The total # of cations is 4.

The net charge of the structure is thus:

# anions:	8 x -1	=	-8
# cations:	4 x +2	=	+8
net charge:		=	0

There are 8 anions and 4 cations, leading to a net charge of zero

Problem 12.18

Determine the number of anions and the number of cations in the corundum structure unit cell, and prove that the structure has a net charge of zero for Al^{3+} and O^{2-} ions.

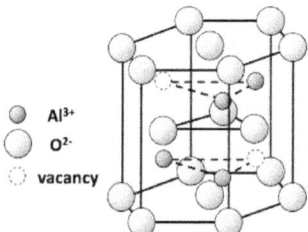

The corundum structure has twelve corner anions each contributing $1/6$ of an anion each and three center anions each contributing 1 anion each. The total # of anions is:

face anions	$2 \times 1/2$	=	1
center anions:	3×1	=	3
corner anions:	$12 \times 1/6$	=	2
Total:		=	6

The corundum structure has four interior cations each contributing a full cation each. The total # of cations is 4.

The net charge of the structure is thus:

# anions:	6×-2	=	-12
# cations:	$4 \times +3$	=	+12
net charge:		=	0

There are 6 anions and 4 cations, leading to a net charge of zero

Problem 12.19

Identify the repeat unit of the following polymer (polyvinylchloride) and write the shorthand.

$$\cdots - CH - CH_2 - CH - CH_2 - CH - CH_2 - CH - CH_2 - \cdots$$
$$||||$$
$$ClClClCl$$

$$\left[CH - CH_2 \atop | \atop Cl \right]_n$$

Problem 12.20

Identify the repeat unit of the following polymer (polystyrene) and write the shorthand structure.

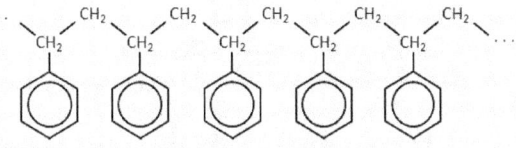

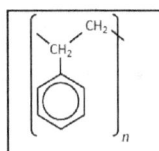

Problem 12.21

Four polymer molecules have molecular weights of 10,000, 20,000, and 30,000, and 50,000 g/mol. Calculate the number average molecular weight the mixed polymer sample.

The number average molecular weight is given as:

$$\overline{M_n} = \frac{\Sigma N_i M_i}{\Sigma N_i}$$

$$\overline{M_n} = \frac{(10,000) + (20,000) + (30,000) + (50,000)}{(4)}$$

$$\overline{M_n} = \frac{110,000}{4}$$

$$\boxed{\overline{M_n} = 27,500 \text{ g/mol}}$$

Problem 12.22

A mixture of polystyrene contains three distinct molecular weights in the following proportions:

$$
\begin{array}{lll}
1\,g & @ & 10{,}000\ g/mol \\
2\,g & @ & 50{,}000\ g/mol \\
2\,g & @ & 100{,}000\ g/mol
\end{array}
$$

Calculate the number average molecular weight of the polystyrene mixture.

The number average molecular weight is given as:

$$\overline{M_n} = \frac{\sum N_i\, M_i}{\sum N_i}$$

$$\overline{M_n} = \frac{(1 \cdot 10{,}000) + (2 \cdot 50{,}000) + (2 \cdot 100{,}000)}{(1 + 2 + 2)}$$

$$\overline{M_n} = \frac{160{,}000}{5}$$

$$\boxed{\overline{M_n} = 32{,}000\ g/mol}$$

Problem 12.23

The atomic structure of two materials are shown below. Which material would you expect to have a clearly-defined melting point?

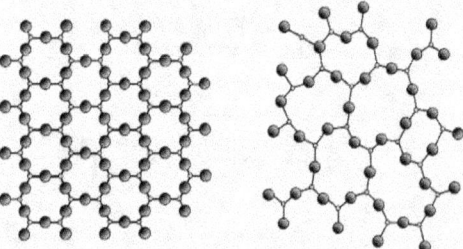

The crystalline structure will have a well-defined melting point.

Problem 12.24

A surfactant is added to a solution that is 70% oil and 30% water. Which of the two micelles below do you expect to form?

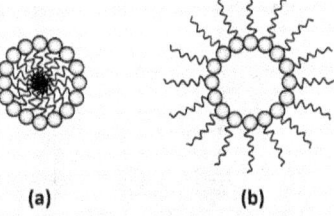

(a)   (b)

The reverse micelle (b) will form. The solution is mostly oil, so the micelles will form with the hydrophobic tails pointed outwards and surrounding the water droplets.

Problem 12.25

The data in the table below was obtained from a tensile test of a metal specimen with a rectangular cross section of 1 cm² in area and a gage length of 5.0 cm. The specimen was not loaded to failure.
   a. Generate a table of stress and strain values (report stress in kPa)
   b. Plot a stress-strain curve

Load (N)	Elongation
0	0
5000	0.00005
11000	0.0001
25000	0.0002
31000	0.00025
32500	0.0005
35000	0.00075
38000	0.001
41000	0.00125
43000	0.0015
45000	0.00175
42000	0.002
35000	0.00225
0	0

We will generate a table of stress and strain values in Excel. The stress is the load divided by cross-sectional area ($\sigma = F/A_o$), and strain is elongation divided by the gauge length ($\varepsilon = $ elongation$/L_o$). We build our table such as:

	A	B	C	D
1	Ao		0.001 m2	
2	Lo		0.05 m	
3				
4				
5				
6	Load (N)	Elongation (m)	Stress (kPa)	Strain
7	0	0	=A7/$B$1/1000	=B7/$B$2
8	5000	0.00005	=A8/$B$1/1000	=B8/$B$2

For a final table:

Load (N)	Elongation (m)	Stress (kPa)	Strain
0	0	0	0
5000	0.00005	5000	0.001
11000	0.0001	11000	0.002
25000	0.0002	25000	0.004
31000	0.00025	31000	0.005
32500	0.0005	32500	0.01
35000	0.00075	35000	0.015
38000	0.001	38000	0.02
41000	0.00125	41000	0.025
43000	0.0015	43000	0.03
45000	0.00175	45000	0.035
42000	0.002	42000	0.04
35000	0.00225	35000	0.045

Plotting the values on a scatter plot connected with curves:

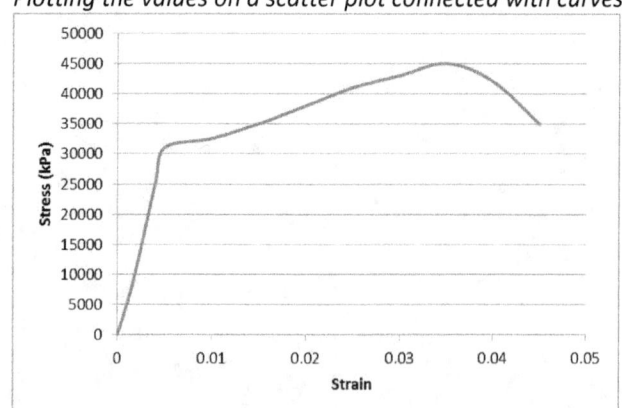

Problem 12.26

A carbon fiber composite is made by adding 90% by volume carbon fibers with elastic modulus of 70 GPa to a polymeric resin with elastic modulus of 5 GPa. The fibers are added such that the composite has a one-dimensional reinforcement: the elastic modulus in the direction of the fibers is equal to the upper limit, and along the width of the fibers is equal to the lower limit. Determine the elastic modulus of the fiber composite in both directions.

The elastic modulus of the composite along fiber orientation (strong direction) is:
$$E_{C,U} = E_M V_M + E_R V_R$$
$$E_{C,U} = (0.1 \cdot 5 \text{ GPa}) + (0.9 \cdot 70 \text{ GPa})$$

$$\boxed{E_{C,U} = 63.5 \text{ GPa}}$$

The elastic modulus of the composite in the weak direction is:
$$E_{C,L} = \frac{E_M E_R}{E_M V_R + E_R V_M}$$
$$E_{C,L} = \frac{(5 \text{ GPa} \cdot 70 \text{ GPa})}{(0.9 \cdot 5 \text{ GPa}) + (0.1 \cdot 70 \text{ GPa})}$$

$$\boxed{E_{C,L} = 30.4 \text{ GPa}}$$

Problem 12.27

Running shoe midsoles absorb energy and decrease the force transferred to the heel. Midsoles are generally made of foam plastic polymer, which compress on impact and expand back to the original shape so it is ready for the next foot strike. The majority of running shoe midsoles are made from ethylene-vinyl acetate (EVA) foam, formed by trapping pockets of air inside of a polymer resin. EVA resin has an elastic modulus of 0.011 GPA and density 0.951 g/cm3. Design the lightest EVA foam possible while maintaining an elastic modulus of at least 1.5 MPa. Report the density of EVA foam in the final design. *Hint: treat the EVA foam as a particle-reinforced composite where air is the reinforcement but provides no structural support.*

The elastic modulus is a summation property, but in this case air provides no support (has a modulus of zero). We can determine the minimum volume fraction of the polymer:
$$E_{C,U} = E_M V_M + E_R V_R$$
$$1.5 \text{ MPa} = 0.011 \text{ GPA}(V_M) + 0 \cdot V_R$$
$$1.5 \times 10^6 \text{ Pa} = 0.011 \times 10^9 \text{ Pa}(V_M)$$
$$\frac{1.5 \times 10^6 \text{ Pa}}{0.011 \times 10^9 \text{ Pa}} = V_M$$
$$V_m = 0.136$$

Now we calculate the density of the resulting composite foam:
$$\rho_{composite} = \rho_M V_M + \rho_R V_R$$
$$\rho_{composite} = \left(0.951 \frac{g}{cm^3}\right) V_M + \left(0.0012 \frac{g}{cm^3}\right)(1 - V_M)$$
$$\rho_{composite} = \left(0.951 \frac{g}{cm^3}\right) 0.136 + \left(0.0012 \frac{g}{cm^3}\right)(0.864) = 0.130 \frac{g}{cm^3}$$

$$\boxed{\text{The density of the final foam will be about } 0.130 \text{ g/cm}^3}$$

Problem 12.28

You have been tasked with selecting the material with which to build a bridge support column. The column will be 10 meters high and designed to support a weight of 18,000 N (a mass of approximately two tons). Given the three available materials below, which is the best choice? Explain.

Material	Compressive Strength (kPa)	Density (kg/m³)	Material Cost ($/kg)
A	80,000	2165	$14.00
B	14,000	2400	$5.20
C	8,200	670	$3.65

Because we need to repeat the same calculations for each material, we will solve this problem using by creating a table in Excel.

The cross-section needed to support the weight is given by:

$$\sigma = \frac{F}{A_0} \rightarrow A_0 = \frac{F}{\sigma}$$

We build our table such as:

	A	B	C	D	E	F	G	H
1	Force		18000 N					
2	Column Height		10 m					
3								
4	Material	Comp. Strength (kPa)	Density (kg/m³)	Cost ($/kg)	min Cross-section (m²)	min Volume (m³)	Mass Needed (kg)	Cost ($)
5	A	80,000	2165	14	=$B$1/B5	=E5*$B$2	=F5*C5	=G5*D5
6	B	14,000	2400	5.2	=$B$1/B6	=E6*$B$2	=F6*C6	=G6*D6
7	C	8,200	670	3.65	=$B$1/B7	=E7*$B$2	=F7*C7	=G7*D7

For final values of:

	A	B	C	D	E	F	G	H
1	Force		18000 N					
2	Column Height		10 m					
3								
4	Material	Comp. Strength (kPa)	Density (kg/m³)	Cost ($/kg)	min Cross-section (m²)	min Volume (m³)	Mass Needed (kg)	Cost ($)
5	A	80,000	2165	14	0.23	2.25	4871.25	$68,197.50
6	B	14,000	2400	5.2	1.29	12.86	30857.14	$160,457.14
7	C	8,200	670	3.65	2.20	21.95	14707.32	$53,681.71

We find that the cheapest option is actually material C.

Chapter 13 Solutions
INDUSTRIAL MANUFACTURING AND OPERATIONS

Problem 13.1
You must produce 600 units in 80 hours to meet a customer order. What is the cycle time required to meet the order?

$$\text{avg cycle time} = \frac{80 \text{ hours}}{600 \text{ units}} \cdot \frac{60 \text{ min}}{\text{hour}} = \frac{4800 \text{ min}}{\text{unit}} = 8 \text{ minutes/unit}$$

$$\boxed{8 \text{ minutes/unit}}$$

Problem 13.2
A bottle washing machine can clean 5 units per hour and has a projected availability of 95%. Determine its weekly capacity.

$$\frac{5 \text{ units}}{\text{hour}} \cdot \frac{168 \text{ hours}}{\text{week}} \cdot 0.95 = 798 \text{ bottles/week}$$

$$\boxed{798 \text{ bottles/week}}$$

Problem 13.3
A laser cutter goes down every day for 10 minutes for focus and alignment, once a week for 8 hours for power supply maintenance, and 4 hours every month for robot calibration. What is the average availability of the laser cutter?

The longest period we can calculate is the monthly availability:

$$A_{month} = \frac{4(168) - 4\left(\frac{10}{60}\right) - 4(8) - 4}{4(168)}$$

$$A_{month} = \frac{672 - 0.67 - 32 - 4}{672}$$

$$A_{month} = 0.945$$

$$\boxed{A_{month} = 0.945}$$

Problem 13.4

Determine the average annual availability of the process tool below:
- Annually down 4 hours for fire & safety audits
- Daily down for 1 hour for spray adjustments
- Weekly down 8 hours for cleaning
- Monthly down 8 hours for robot greasing
- Quarterly down 8 hours for robot maintenance

365 days/year · 24 hrs/day = 8760 hrs/year

$$A_{year} = \frac{8760 - 365(1) - 52(8) - 12(8) - 4(8) - 4}{8760}$$

$$A_{year} = \frac{8760 - 365 - 416 - 86 - 32 - 4}{8760} = 0.897$$

$$\boxed{A_{year} = 0.897}$$

Problem 13.5

Determine the cycle times for the operations below, and determine the hourly throughput. Assume each operation has enough buffer capacity to start each run immediately.

a. A polisher processes one unit at a time. Each unit takes 3.5 minutes
b. An oven can bake 50 loaves of bread at a time. Baking takes 90 minutes.
c. A washer can clean 10 plates at a time at 20 minutes per run. The store has two washers.

a) Cycle time = 3.5 min

$$\text{throughput} = \frac{1 \text{ unit}}{3.5 \text{ mins}} \cdot \frac{60 \text{ mins}}{\text{hour}} = 17.14 \text{ units/hour}$$

$$\boxed{CT = 3.5 \text{ mins, TP} = 17 \text{ units/hr}}$$

b) Cycle time = 90 min

$$\text{throughput} = \frac{1 \text{ batch}}{90 \text{ mins}} \cdot \frac{60 \text{ mins}}{\text{hour}} \cdot \frac{50 \text{ loaves}}{\text{batch}} = 33.3 \text{ units/hour}$$

$$\boxed{CT = 90 \text{ mins, TP} = 33 \text{ loaves/hr}}$$

c) Cycle time = 20 min

$$\text{throughput} = \frac{10 \text{ plates}}{20 \text{ mins}} \cdot \frac{60 \text{ mins}}{\text{hour}} \cdot 2 \text{ tools} = 60 \text{ units/hour}$$

$$\boxed{CT = 90 \text{ mins, TP} = 60 \text{ units/hr}}$$

Problem 13.6

Given the process below, determine the cycle time of:
   a. Operation B
   b. Operation C
   c. The entire process

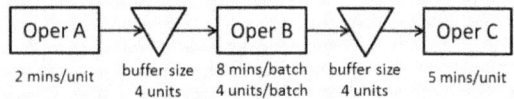

Operation B has a 4 unit buffer, and processes 4 units at a time. That means that:

Unit	Process Time	Wait Time	Total Cycle Time
4	8 min	0 min	8 min
3	8 min	2 min	10 min
2	8 min	4 min	12 min
1	85 min	6 min	14 min
		Total	44 min

$$\text{cycle time} = \frac{44}{4} = 11 \text{ mins}$$

Oper B cycle time is 11 mins

At operation C, the processing time is 5 minutes but a part must sit in the buffer for 20 minutes (4 · 5 mins). The average cycle time for operation C is thus:

$$\text{cycle time} = \text{processing time} + \text{wait time} = 5 \text{ mins} + 20 \text{ mins}$$
$$\text{cycle time} = 25 \text{ mins}$$

Oper C cycle time is 25 mins

The cycle time for the full process is the sum of both operations:

$$\text{process cycle time} = \text{CT oper A} + \text{CT oper B} + \text{CT oper C} = 2 \text{ mins} + 11 \text{ mins} + 25 \text{ mins}$$
$$\text{process cycle time} = 38 \text{ mins}$$

Process cycle time is 38 mins

Problem 13.7
A restaurant has 25 tables and on average diners spend 45 minutes at a table.
    a. What is the average cycle time?
    b. How many parties can be served each hour?
    c. If you came in and were 6th on the waiting list, how long would you expect to wait to be seated?

a)
$$\frac{45 \text{ min}}{25 \text{ tables}} = 1.8 \text{ min/table}$$

$$\boxed{1.8 \text{ min/table}}$$

b)
$$\frac{60 \text{ min}}{1.8 \text{ min/table}} = 33 \text{ tables}$$

$$\boxed{33 \text{ tables}}$$

c) Sixth place on the waiting list means:
$$6 \cdot 1.8 \text{ min/table} = 10.8 \text{ minutes}$$

$$\boxed{\approx \text{about 11 minutes}}$$

Problem 13.8
Given the factory setup in the table below, determine:
    a. The bottleneck
    b. Cycle time for the entire process
    c. Overall process throughput

operation	# of machines	process time	station rate
1	1	2 hr	12 jobs/day
2	2	5 hr	9.6 jobs/day
3	6	10 hr	14.4 jobs/day
4	2	3 hr	16 jobs/day

a) The process bottleneck is operation 2, with the lowest throughput.

b) CT = 20 hours

c) The process throughput is equal to the bottleneck, so 9.6 jobs/day.

Chapter 13 | Industrial Manufacturing and Operations

Problem 13.9

A mobile phone manufacturer has a daily throughput of 1250 phones and a current inventory of 8000 phones. A customer puts in an order for 5000 phones. How long will it take to fulfill the order?

We first find the MLT:

$$\text{MLT} = \frac{\text{WIP}}{\text{throughput}}$$

$$\text{MLT} = \frac{8{,}000 \text{ phones}}{1250 \text{ phones/day}}$$

$$\text{MLT} = 6.4 \text{ days}$$

Now we can calculate the OFT:

$$\text{OFT} = 6.4 \text{ days} + \frac{5000 \text{ phones}}{1250 \text{ phones/day}}$$

$$\text{OFT} = 6.4 \text{ days} + 4 \text{ days} = 10.4 \text{ days}$$

10.4 days

Problem 13.10

A company sells 10,000 parts per month, and has a warehouse that can hold 15,000 parts. What is the average duration a part spends in the warehouse?

The time spent in the warehouse is essentially the MLT of a unit through the warehouse:

$$\text{MLT} = \frac{\text{WIP}}{\text{throughput}}$$

$$\text{MLT} = \frac{15{,}000 \text{ parts}}{10{,}000 \text{ parts/wk}}$$

$$\text{MLT} = 1.5 \text{ weeks}$$

1.5 weeks in the warehouse

Problem 13.11
A bakery which operates 24x7 has a design capacity of 1200 rolls/hour and an effective capacity of 175,000 rolls per week. If last week's actual production was 148,000 rolls, determine the utilization and efficiency of the bakery.

First calculate the weekly design capacity:
Design capacity = 24 hours/day · 7 days/week · 1200 rolls/hour = 201,600 rolls/week

Can now calculate the design capacity and the efficiency of the bakery:

$$\text{utilization} = \frac{\text{actual output}}{\text{design capacity}} = \frac{148{,}000 \text{ rolls}}{201{,}600 \text{ rolls}} = 73.6\%$$

$$\text{efficiency} = \frac{\text{actual output}}{\text{effective capacity}} = \frac{148{,}000 \text{ rolls}}{175{,}000 \text{ rolls}} = 84.6\%$$

> **Utilization = 73.6%**
> **Efficiency = 84.6%**

Problem 13.12
Your factory is running at 75% utilization and 75% efficiency. Where should you focus your improvement efforts? Why?

We should first try and improve efficiency, because it will generally easiest, quickest, and cheapest.

Problem 13.13

A grain milling machine can crush 100 lb of grain per hour and has an average availability of 0.95. Over the duration of one week, the machine processed a total of 12,000 lbs of grain. Determine the design capacity, effective capacity, utilization, and efficiency of the mill.

The design capacity is the maximum output ignoring downtime:

$$\text{throughput} = \frac{100 \text{ lb}}{\text{hour}} \cdot \frac{168 \text{ hour}}{\text{week}} = 16{,}800 \text{ lbs/week}$$

Design capacity = 16,800 lbs/week

The effective capacity is the maximum output taking availability into account:

$$\text{throughput} = \frac{100 \text{ lb}}{\text{hour}} \cdot \frac{168 \text{ hour}}{\text{week}} \cdot 0.95 = 15{,}960 \text{ lbs/week}$$

Effective capacity = 15,960 lbs/week

Can now calculate the design capacity and the efficiency of the bakery:

$$\text{utilization} = \frac{12{,}000 \text{ lbs}}{16{,}800 \text{ lbs}} = 71.4\%$$

$$\text{efficiency} = \frac{12{,}000 \text{ lbs}}{15{,}960 \text{ lbs}} = 75.2\%$$

Utilization = 71.4%
Efficiency = 75.2%

Problem 13.14
In the assembly plant structure below, where is the bottleneck for product C?

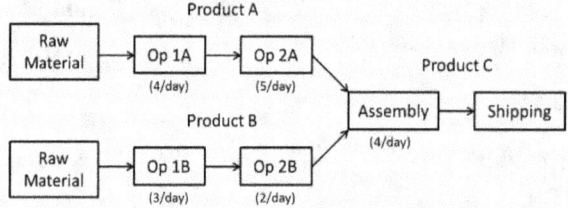

Operation 2B has the lowest throughput of product B, and also limits the whole process.

Problem 13.15
In the V-plant production scheme shown below, where is the bottleneck for product A? For product B?

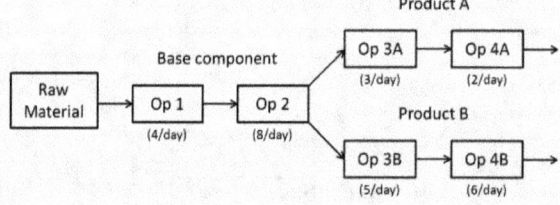

Product A is limited by operation 4A, but product B is limited by operation 1.

Problem 13.16

A factory that makes frozen vegetables has the production process shown below. Vegetables enter the factory and go through each operation (washing, packaging, freezing) in sequence; there are 3 washers, two 2 packagers, and one freezer. Each washer can handle 25 pounds per hour, each packager can handle 30 pounds per hour, and the freezer can handle 80 pounds per hour.

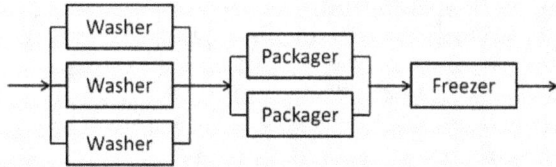

d. What is the limiting step?
e. Assuming a 12-hour work day, find the weekly capacity of the factory to make 1-lb bags.
f. If the factory were to add a third packager, what would be the new line capacity?

To find the bottleneck, we must calculate the throughput of each operation:

Washing = 25 lbs/hr · 3 tools = 75 lbs/hr
Packaging = 30 lbs/hr · 2 tools = 60 lbs/hr
Freezer = 80 lbs/hr · 1 tool = 80 lbs/hr

Packaging is the bottleneck

Process capacity is the capacity of the bottleneck:

$$\frac{60 \text{ lbs}}{\text{hour}} \cdot \frac{\text{bag}}{1 \text{ lb}} \cdot \frac{12 \text{ hours}}{\text{day}} \cdot \frac{5 \text{ days}}{\text{week}} = 3600 \text{ bags/week}$$

3600 bags/wk

A third packager shifts the bottleneck to the freezer, so we calculate based on the capacity of the new bottlneck:

$$\frac{80 \text{ lbs}}{\text{hour}} \cdot \frac{\text{bag}}{1 \text{ lb}} \cdot \frac{12 \text{ hours}}{\text{day}} \cdot \frac{5 \text{ days}}{\text{week}} = 3600 \text{ bags/week}$$

4800 bags/wk

Problem 13.17

A pizza shop has the production process shown below. There are four sequential operations (dough, toppings, baking, boxing), and there are two ovens. The crew can make a batch of 10 pizza doughs every 15 minutes, the toppings crew can top a pizza every 2 minutes, and the boxer can do a pizza every minute. A single oven can hold 5 pizzas, and a batch of pizzas must bake for 20 minutes.

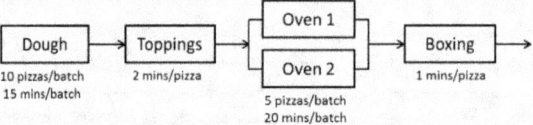

a. What is the cycle time for a single pizza?
b. What is the lead time for a batch of 10 pizzas?
c. Assuming the shop is open for 6 hours a night, what is the daily capacity of the pizza shop?

The cycle time for a single pizza through the line is:

$$= 15 \text{ min} + 2 \text{ min} + 20 \text{ min} + 1 \text{ min} = 38 \text{ mins}$$

Single pizza = 38 mins

The lead time for a batch of 10 pizzas through the line is:

$$= 15 \text{ min/batch} + 2 \text{ min/pizza} \cdot 10 \text{ pizzas} + 20 \text{ min} + 1 \text{ min} \cdot 10 \text{ pizzas}$$
$$= 15 + 20 + 20 + 10 = 65 \text{ minutes}$$

10 pizzas = 65 mins

The process bottleneck is the baking operation. The ovens can produce 10 pizzas every 20 minutes, so:

$$\frac{10 \text{ pizzas}}{20 \text{ mins}} \cdot \frac{6 \text{ hours}}{day} \cdot \frac{60 \text{ min}}{hour} = 180 \text{ pizzas/night}$$

180 pizzas/night

Problem 13.18

A factory's production target is 1000 units per week. If a single cleaning machine takes 30 minutes per unit, how many cleaning machines should be purchased if the cleaning machine availability is 90%?

The throughput of a single tool is:

$$\frac{\text{units}}{0.5 \text{ hour}} \cdot \frac{168 \text{ hours}}{\text{week}} \cdot 0.90 = 75.6 \text{ units/wk}$$

The number of tools needed to meet capacity:

$$\frac{1000 \text{ units/wk}}{75.6 \frac{\text{units}}{\text{wk}} \text{ per tool}} = 13.22 \text{ tools} = 14 \text{ tools}$$

14 tools

Problem 13.19

Consider the production line below. If workstation B must go down for 8 hours of maintenance, how large should the upstream and downstream buffers be such that workstations A and C can run uninterrupted through the downtime?

→ A → ▽ → B → ▽ → C →
  30 mins/widget   1 batch/hour
                        3 widgets/batch

Upstream buffer size:

$$\frac{8 \text{ hours}}{30 \text{ min/widget}} \cdot \frac{60 \text{ min}}{\text{hour}} = 16 \text{ widgets}$$

Downstream buffer size:

$$\frac{8 \text{ hours}}{1 \text{ hour/batch}} \cdot \frac{3 \text{ widgets}}{\text{batch}} = 24 \text{ widgets}$$

Upstream buffer should be 16 widgets, downstream should be 24 widgets

Problem 13.20

Given the candy packaging process shown, how large should the buffer at boxing be to ensure smooth operations?

```
🍬🍬 → [BAGGING] → CANDY → [BOXING] → [CANDY]
        1 bag/minute        16 bags → box
                            1 box/minutes
```

We should set the buffer inventory at 16 units.

Problem 13.21

In a semiconductor manufacturing process, wafers are sent to a wet etch tool fifteen three times throughout the process. The wet bench has a weekly availability of 90% and can clean a wafer every 6 minutes. How many activities can the wet bench perform each week? What is the weekly process output?

The cutting machine can process one part every 5 minutes:

$$\frac{1 \text{ part}}{6 \text{ mins}} \cdot \frac{10{,}080 \text{ mins}}{\text{week}} \cdot 0.90 = 1680 \text{ parts/week}$$

This number reflects the total activities, and because three activities happen for each final output, the process output is one-third, or 604 parts/week.

> **Tool capacity is 1680 activities per week**
> **Process capacity is 112 parts/week**

Problem 13.22

The table below gives the process flow for a single metal layer in a semiconductor manufacturing process. To completely build the circuitry onto a computer chip, this entire process is repeated 8 times. All processing tools are held to an availability of 90%, except for lithography which is 85%. How many of each type of equipment is needed if the semiconductor fab has an output target of 1000 wafers/wk?

op#	activity	process tool	process time
1	oxide deposition	furnace	180 min/100 wafer batch
2	photoresist	spin coater	3 min/wafer
3	patterning	lithography	4.5 min/wafer
4	dry etch	plasma etcher	5 min/wafer
5	wet cleans 1	wet bench	2 min/wafer
6	copper deposition	electroplater	3.5 min/wafer
7	copper polish	polisher	1.5 min/wafer
8	wet cleans 2	wet bench	2 min/wafer
9	inspection	metrology	1 min/wafer

Because the process is repeated 8 times, every single operation must be treated as re-entrant. Also, the wet etch module is repeated twice, and thus has a reentrancy of 16 times.

The oxide deposition furnace:

$$\text{furnace throughput} = \frac{1 \text{ batch}}{180 \text{ mins}} \cdot \frac{100 \text{ wafers}}{\text{batch}} \cdot \frac{10{,}080 \text{ mins}}{\text{week}} \cdot 0.90 = 5{,}040 \text{ activities/week}$$

$$\text{\# furnaces} = \frac{1000 \frac{\text{wafers}}{\text{week}} \cdot 8 \frac{\text{activities}}{\text{wafer}}}{5040 \frac{\text{activities}}{\text{week} \cdot \text{machine}}} = 1.59 \rightarrow 2 \text{ furnaces}$$

The spin coater:

$$\text{spin coater throughput} = \frac{1 \text{ wafers}}{3 \text{ mins}} \cdot \frac{10{,}080 \text{ mins}}{\text{week}} \cdot 0.90 = 3{,}024 \text{ activities/week}$$

$$\text{\# spin coaters} = \frac{1000 \frac{\text{wafers}}{\text{week}} \cdot 8 \frac{\text{activities}}{\text{wafer}}}{3024 \frac{\text{activities}}{\text{week} \cdot \text{machine}}} = 2.65 \rightarrow 3 \text{ spin coaters}$$

The lithography tool:

$$\text{litho throughput} = \frac{1 \text{ wafers}}{4.5 \text{ mins}} \cdot \frac{10{,}080 \text{ mins}}{\text{week}} \cdot 0.85 = 1904 \text{ activities/week}$$

$$\text{\# litho tools} = \frac{1000 \frac{\text{wafers}}{\text{week}} \cdot 8 \frac{\text{activities}}{\text{wafer}}}{1904 \frac{\text{activities}}{\text{week} \cdot \text{machine}}} = 4.20 \rightarrow 5 \text{ litho tools}$$

The plasma etcher:

$$\text{plasma etcher throughput} = \frac{1 \text{ wafers}}{5 \text{ mins}} \cdot \frac{10{,}080 \text{ mins}}{\text{week}} \cdot 0.90 = 1814.4 \text{ activities/week}$$

$$\text{\# plasma etchers} = \frac{1000 \frac{\text{wafers}}{\text{week}} \cdot 8 \frac{\text{activities}}{\text{wafer}}}{1814.4 \frac{\text{activities}}{\text{week} \cdot \text{machine}}} = 4.41 \rightarrow 5 \text{ plasma etchers}$$

The electroplater:

$$\text{plater throughput} = \frac{1 \text{ wafers}}{3.5 \text{ mins}} \cdot \frac{10{,}080 \text{ mins}}{\text{week}} \cdot 0.90 = 2{,}592 \text{ activities/week}$$

$$\# \text{ platers} = \frac{1000 \frac{\text{wafers}}{\text{week}} \cdot 8 \frac{\text{activities}}{\text{wafer}}}{2592 \frac{\text{activities}}{\text{week} \cdot \text{machine}}} = 3.09 \rightarrow 4 \text{ electroplaters}$$

The polisher:

$$\text{polisher throughput} = \frac{1 \text{ wafers}}{1.5 \text{ mins}} \cdot \frac{10{,}080 \text{ mins}}{\text{week}} \cdot 0.90 = 6{,}048 \text{ activities/week}$$

$$\# \text{ polishers} = \frac{1000 \frac{\text{wafers}}{\text{week}} \cdot 8 \frac{\text{activities}}{\text{wafer}}}{6048 \frac{\text{activities}}{\text{week} \cdot \text{machine}}} = 1.32 \rightarrow 2 \text{ polishers}$$

The inspection tool:

$$\text{metrology throughput} = \frac{1 \text{ wafers}}{1 \text{ mins}} \cdot \frac{10{,}080 \text{ mins}}{\text{week}} \cdot 0.90 = 9{,}072 \text{ activities/week}$$

$$\# \text{ metrology tools} = \frac{1000 \frac{\text{wafers}}{\text{week}} \cdot 8 \frac{\text{activities}}{\text{wafer}}}{9072 \frac{\text{activities}}{\text{week} \cdot \text{machine}}} = 0.88 \rightarrow 1 \text{ metrology tool}$$

The wet bench:

$$\text{wet bench throughput} = \frac{1 \text{ wafers}}{2 \text{ mins}} \cdot \frac{10{,}080 \text{ mins}}{\text{week}} \cdot 0.90 = 4{,}536 \text{ activities/week}$$

$$\# \text{ wet benches} = \frac{1000 \frac{\text{wafers}}{\text{week}} \cdot 16 \frac{\text{activities}}{\text{wafer}}}{4536 \frac{\text{activities}}{\text{week} \cdot \text{machine}}} = 3.52 \rightarrow 4 \text{ wet benches}$$

The factory needs:
- **1 metrology tool**
- **2 furnaces**
- **2 polishers**
- **3 spin coaters**
- **4 electroplaters**
- **4 wet benches**
- **5 lithography tools**
- **5 plasma etchers**

Problem 13.23

Consider the laundry process shown. If you need to do three loads of laundry, one load will be forced to process alone at the dryer. Would it be more efficient for the first load to process alone, or for the last load to process alone?

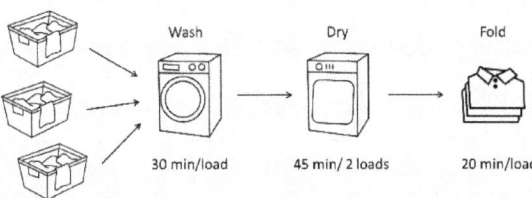

If the first load is put into the dryer as soon as it is finished, the total duration for all three loads is 175 mins:

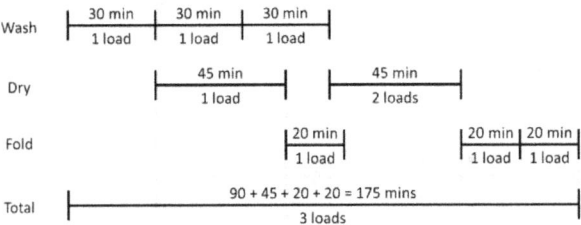

If we wait for two loads to be washed before we dry, the total duration for all three loads is 170 mins:

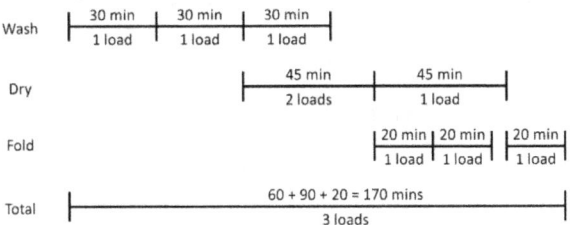

The difference is the utilization of the folding process. Even though there is more idle time in-between runs, we are able to more efficiently use the folding station (as it can process two loads while the drier is running the last load). **We save 5 minutes overall by processing the last load alone.**

Solutions
ANSWERS TO COMPREHENSIVE LEARNING PROBLEMS

Chapter 5 Comprehensive Learning
RESISTANCE OF A BOLT

For current to flow from end-to-end through the bolt, it passes through the head and the body/shaft in succession. Although they are made of the same material (same resistivity), each region has a different resistance because of their dimensions. We thus treat the bolt as two resistors in series:

head   body

Therefore: $R_{bolt} = R_{head} + R_{body}$

> **Solution steps:**
> 1 – Determine the resistance of the bolt body
> 2 – Determine the resistance of bolt head
> 3 – Calculate overall resistance of the entire bolt

Let us first solve for the resistance of the bolt body.
The cross-sectional area of the body is the area of a circle. The diameter is 0.008 m, which means:

$A = \pi r^2$

$A = \pi \left(\dfrac{d}{2}\right)^2$

$A = \pi \left(\dfrac{0.008 \text{ m}}{2}\right)^2$

$A = 5.03 \times 10^{-5} \text{ m}^2$

We can now easily calculate the resistance of the body:

$R_{body} = \rho \dfrac{L}{A_C}$

$R_{body} = (6.90 \times 10^{-7} \; \Omega m) \dfrac{0.15 \text{ m}}{5.03 \times 10^{-5} \text{ m}^2}$

$R_{body} = 2.06 \times 10^{-3} \; \Omega$

We next must solve for the resistance of the bolt head.
The cross sectional area of the bolt heat is a regular hexagon. The area of a regular hexagon is:

$A = \dfrac{3\sqrt{3}}{2} d^2$

$A = \dfrac{3\sqrt{3}}{2} (0.01 \text{m})^2$

$A = 2.60 \times 10^{-4} \text{ m}^2$

We can now easily calculate the resistance of the head:

$$R_{body} = \rho \frac{L}{A_C}$$

$$R_{body} = (6.90 \times 10^{-7} \text{ }\Omega\text{m}) \frac{0.007 \text{ m}}{2.60 \times 10^{-4} \text{ m}^2}$$

R_{body} = 1.85 x 10^{-5} Ω

We can now solve for the total resistance of the entire bolt.

$R_{bolt} = R_{head} + R_{body}$

$R_{bolt} = 1.85 \times 10^{-5}$ Ω + 2.06 x 10^{-3} Ω

$R_{bolt} = 2.0785 \times 10^{-3}$ Ω

$R_{bolt} = 2.0785 \times 10^{-3}$ Ω

Chapter 6 Comprehensive Learning
POT ON A STOVE

> **Solution steps:**
> 1 – Unit conversion to get:
>   - Mass of H_2O
>   - Time in seconds
> 2 – Calculate energy required to heat the water to boiling
> 3 – Calculate power required to do so in the allotted time
> 4 – Calculate the current needed to generate the necessary power

First, we need to find the weight of 1 gallon of water (density = 1 g/cm³), and the time in seconds:

$$\frac{1 \text{ gallon H}_2\text{O}}{} \left| \frac{3.785 \text{ L}}{1 \text{ gal}} \right| \frac{1000 \text{ mL}}{\text{L}} \left| \frac{1 \text{ cm}^3}{\text{mL}} \right| \frac{1 \text{ g}}{\text{cm}^3}$$

$$\frac{1 \text{ \sout{gallon} H}_2\text{O}}{} \left| \frac{3.785 \text{ \sout{L}}}{1 \text{ \sout{gal}}} \right| \frac{1000 \text{ \sout{mL}}}{\text{\sout{L}}} \left| \frac{1 \text{ \sout{cm}}^3}{\text{\sout{mL}}} \right| \frac{1 \text{ g}}{\text{\sout{cm}}^3}$$

1 gal H_2O = 3785 g H_2O

10 minutes = 600 seconds

Second, we need to find the energy required to heat that mass of water by 75 °C:
$Q = m \Delta T \, c$
$Q = (3785 \text{ g})(373 \text{ K} - 298 \text{ K})(4.18 \text{ J/g K})$
$Q = (3785 \text{ g})(75 \text{ K})(4.18 \text{ J/gK})$
$Q = 1.19 \times 10^6 \text{ J}$

Third, the power is simply the energy per time:
$P = Q/t$
$P = 1.19 \times 10^6 \text{ J} / 600 \text{ s}$
$P = 1.98 \times 10^3 \text{ Watts}$

Last, we calculate the current:
$P = I V \rightarrow I = P/V$
$I = 1.98 \times 10^3 \text{ Watts} / 240 \text{ V}$

$\boxed{I = 8.23 \text{ A}}$

Chapter 7 Comprehensive Learning
WATER TOWER PUMP

> **Solution steps:**
> 1 – Unit conversion to get Mass of H_2O
> 2 – Calculate energy required to lift the water
> 3 – Calculate overall energy required to the pump
> 4 – Calculate the time needed given the power and current supplied

First, we need to find the weight of 1 gallon of water (density = 1 g/cm³), and the time in seconds:

$$\frac{10000 \text{ gallon } H_2O}{} \cdot \frac{0.003785 \text{ m}^3}{1 \text{ gal}} \cdot \frac{1000 \text{ kg}}{\text{m}^3}$$

10000 gal H_2O = 37850kg H_2O

Second, we need to find the energy required to lift that mass of water by 20m:
W = Force • Distance
W = (9.8 m/s²) (37850 kg) (20 m)
W = 7.4186x 10^6 J

Third, the overall energy consumed by the pump:
W_T = W/0.7
W_T = 7.4186 x 10^6 J /0.7
W_T = 1.0598 x 10^7 J

Last, we calculate the time required:
P = I V t → t = Q / I V
t = 1.0598 x 10^7 J/ (240 V)(60 A)
t = 735.97 seconds

> t = 12.27minutes

Chapter 8 Comprehensive Learning
HEAT EXCHANGER

a) You open valve #1 from 50% to 75%. What are the effects to $T_{H,OUT}$ and $T_{C,OUT}$?

When valve #1 opens, the flow rate of the hot stream increases. $T_{H,OUT}$ and $T_{C,OUT}$ will both increase.

b) You close valve #2 from 50% to 25%. What are the effects to $T_{H,OUT}$ and $T_{C,OUT}$?

When valve #2 closes, the flow rate of the cold stream decreases. $T_{H,OUT}$ and $T_{C,OUT}$ will both increase.

c) The inlet temperature of the hot stream ($T_{H,IN}$) increases. What action should you take such that the outlet stream temperature remains unchanged? When you take this action, what happens to $T_{C,OUT}$?

The proper action is to open valve #2, increasing the flow rate of cooling water. This will allow for more total heat transfer to the cooling water stream. $T_{C,OUT}$ will initially increase, but as valve #2 is opened it will drop.

d) The inlet temperature of the cold stream ($T_{C,IN}$) decreases. What action should you take such that the outlet stream temperature remains unchanged? When you take this action, what happens to $T_{C,OUT}$?

A decrease in the cooling water supply temperature will require you to increase the flow rate of cooling water by opening valve #2. $T_{C,OUT}$ will initially increase, but as valve #2 is opened it will drop.

Bonus: You open valve #2 from 90% to 100%, but there are no changes in the system. What has happened?

Opening the valve should increase the flow of cooling water. However, because no changes are observed, the flow must not have changed. Thus, we must have hit the maximum supply flow of the cooling water.

Chapter 12 Comprehensive Learning
BOLT STRENGTH

> **Solution steps:**
> 1 – Determine the yield point of the material
> 2 – Calculate the mass that a single bolt can support
> 3 – Determine how many bolts will be needed to support the pipe

We must ensure that the stress on the material remains below the yield point. From the stress-strain diagram, the yield point of the material appears to be approximately 23 MPa:

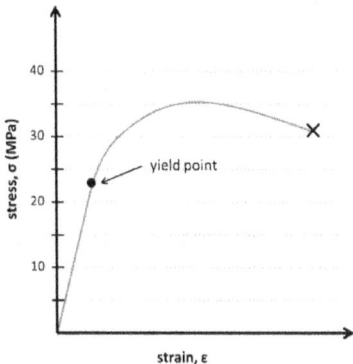

The cross-sectional area of the bolt is:
$A = \pi (d/2)^2$
$A = \pi (0.01m/2)^2$
$A = \pi (0.005m/2)^2$
$A = 7.85 \times 10^{-5} \, m^2$

Thus, the maximum mass that can be supported at the yield point is:
$\sigma = \dfrac{F}{A_o}$
$F = \sigma \cdot A_o$
$m \cdot a = 23 \, MPa \cdot 7.85 \times 10^{-5} \, m^2$
$m \cdot 9.8 \, m/s^2 = 23 \times 10^6 \, N/m^2 \cdot 7.85 \times 10^{-5} \, m^2$
$m \cdot 9.8 \, m/s^2 = 1805.5 \, N$
$m = 184.2 \, kg$

So, a single bolt can support 184 kg. To find the total number of bolts needed to support the 1 ton pipe:

1 ton	907.185 kg	1 bolt	= 4.925 bolts
	ton	184.2 kg	

> **We to use need 5 bolts to safely support the pipe**

Chapter 13 Comprehensive Learning
INTERNET SERVICE PROVIDER

> **Solution steps:**
> 1 – Determine the number of components needed to meet maximum demand
> 2 – Determine the number of each component needed to be connected in parallel

In order to meet peak demand, we need:

$$= \frac{200 \frac{\text{Mbps}}{\text{user}} \cdot 1000 \text{ users}}{1000 \frac{\text{Mbps}}{\text{switch}}}$$
$$= 200 \text{ switches}$$

$$= \frac{200 \frac{\text{Mbps}}{\text{user}} \cdot 1000 \text{ users}}{1000 \frac{\text{Mbps}}{\text{switch}}}$$
$$= 2.66 \text{ servers} \rightarrow 3 \text{ servers}$$

This gives the minimum number of each component that must be simultaneously working in order to meet demand. These components must thus be considered to be in series, which will have low reliability. The best way to increase the reliability is to connect in low-level redundancy.

To determine how many of each component must be in parallel for each group, we go to the equation for a low-level redundant system:

$$R_S = [1 - (1 - R_C)^{\text{\# in parallel}}]^{\text{\# sets in series}}$$

Solving for each component:

$$R_{\text{switches}} = [1 - (1 - R_{\text{switch}})^{\text{\# in parallel}}]^{\text{\# sets in series}}$$
$$R_{\text{switches}} = [1 - (1 - 0.98)^{\text{\# in parallel}}]^{200}$$

\# in parallel = 3.12, so *4 switches in parallel*

$$R_{\text{servers}} = [1 - (1 - R_{\text{server}})^{\text{\# in parallel}}]^{\text{\# sets in series}}$$
$$R_{\text{servers}} = [1 - (1 - 0.95)^{\text{\# in parallel}}]^{3}$$

\# in parallel = 2.67, so *3 servers in parallel*

Now drawing the RBD for each:

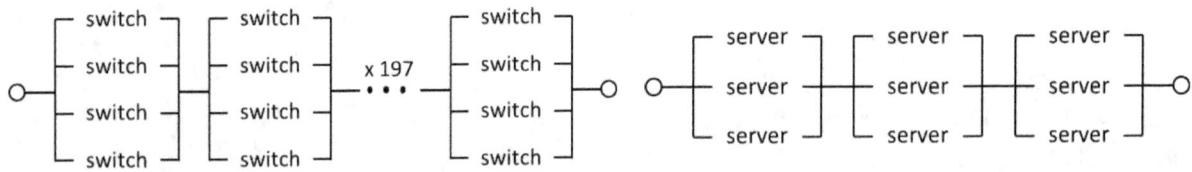

200 sets of 4 parallel switches   **3 sets of 3 parallel servers**

www.ingramcontent.com/pod-product-compliance
Lightning Source LLC
Chambersburg PA
CBHW080009210526
45170CB00015B/1954